An Economic Assessment of Banana Genetic Improvement and Innovation in the Lake Victoria Region of Uganda and Tanzania

Edited by Melinda Smale and Wilberforce Kateera Tushemereirwe

Principal authors in alphabetical order:
Pamela Nahamya Abodi, Fredrick Bagamba,
Mgenzi Said Ramadhan Byabachwezi, Svetlana Edmeades, Robert Kalyebara,
Deborah Karamura, Enid Katungi, Enoch Mutebi Kikulwe, Jackson M. Nkuba,
Melinda Smale, Wilberforce Kateera Tushemereirwe, and Stanley Wood

RESEARCH REPORT 155

INTERNATIONAL FOOD POLICY RESEARCH INSTITUTE
sustainable solutions for ending hunger and poverty
IFPRI®

IN COLLABORATION WITH
AGRICULTURAL RESEARCH AND DEVELOPMENT INSTITUTE
BIOVERSITY INTERNATIONAL
NATIONAL AGRICULTURAL RESEARCH ORGANIZATION

International Food Policy Research Institute
2033 K Street, NW
Washington, D.C. 20006-1002, U.S.A.
Telephone +1-202-862-5600
www.ifpri.org

DOI: 10.2499/9780896291645RR155

Library of Congress Cataloging-in-Publication Data

An economic assessment of banana genetic improvement and innovation in the Lake Victoria region of Uganda and Tanzania / M. Smale and W. Tushemereirwe, eds.
p. cm. — (IFPRI research report ; 155)
Includes bibliographical references.
ISBN-13: 978-0-89629-164-5 (alk. paper)
ISBN-10: 0-89629-164-2 (alk. paper)
1. Bananas—Genetic engineering—Economic aspects—Uganda. 2. Bananas—Genetic engineering—Economic aspects—Tanzania. I. Smale, Melinda. II. Tushemereirwe, W. III. International Food Policy Research Institute. IV. Series: Research report (International Food Policy Research Institute) ; 155.
SB379.B2.E245 2007
338.1'7477209676—dc22 2007021387

Contents

Tables

Figures

Foreword

This research report highlights the findings from a set of studies undertaken by the International Food Policy Research Institute, along with several national and international research institutions, to assess the economic impact of improved cultivars and management practices on smallholder farmers in the Lake Victoria Region of Uganda and Tanzania—an area where the cooking banana is both economically and culturally important. Genetic transformation is a promising alternative for improving the resistance of banana plants to the pests and diseases that cause serious economic losses, because bananas, unlike rice, wheat, and maize, are difficult to improve through conventional breeding techniques.

The team of researchers posed three broad questions: What is the current level of adoption of improved cultivars and management practices, given the constraints to banana production and marketing? What are the prospects that banana growers will adopt cultivars with transgenic resistance to pests and diseases, given existing constraints? What is the potential impact on the banana industry of a range of genetic technologies now under development by national researchers?

Findings confirm that smallholder farmers value the crop traits targeted for introduction by current bioengineering efforts, and that transgenic bananas have the potential to benefit the poor. The focal role of farmers in developing and promoting planting material systems and the influence of social capital on technology adoption are both demonstrated by the baseline data, with implications for the design of extension systems.

The research summarized in this report is new in several respects. First, it documents the uptake of newly released banana hybrids and other recommended banana practices on semi-subsistence smallholder farms in East Africa. Second, a complete taxonomy of distinct banana cultivars grown in the region has been developed. Third, the research puts existing knowledge into perspective with advanced social science methods.

The findings and recommendations resulting from this study will be useful for policy decisions in the region, contributing initially to technological development and dissemination, and ultimately to increased crop productivity.

Joachim von Braun
Director General, IFPRI

Acknowledgments

The International Food Policy Research Institute (IFPRI) undertook the research presented in this report in partnership with the International Plant Genetic Research Institute (IPGRI), now Bioversity International, and an array of individuals and institutional partners—the initiator being IPGRI's International Network for Improvement of Banana and Plantains (INIBAP).

The Agricultural Research and Development Institute (ARDI), Maruku, led a research component in Kagera Region, Tanzania. ARDI staff built on the experiences of a 5-year project, titled Propagation and Diffusion of Superior Banana Cultivars, which was launched by the Kagera Community Development Programme (KCDP) in collaboration with the Belgian Technical Cooperation and Tanzanian authorities. In Uganda, the research was led by staff at the National Banana Research Programme (NBRP) of the National Agricultural Resource Organization (NARO). In Uganda, elite endemic cooking cultivars have been propagated through tissue culture, and NBRP has promoted the use of best management practices to combat banana pests and diseases. Scientific staff from the International Institute for Tropical Agriculture (IITA) also contributed important information and critical review at key junctures. Doctoral and masters students conducted the in-depth studies from which the analyses summarized here are drawn. They were advised by professors at Makerere University, Kampala, Uganda; Sokoine University, Sokoine, Tanzania; University of Pretoria, Pretoria, South Africa; Wageningen University, Wageningen, The Netherlands; and North Carolina State University, Raleigh, North Carolina, United States.

The U.S. Agency for International Development (USAID) funded the research and offered technical guidance. The Rockefeller Foundation and the E.U. also financed research by doctoral students. Funding was provided through the response program of Wageningen University.

The authors particularly acknowledge the support and expertise of Eldad Karamura and his staff at INIBAP in Kampala; the former director of INIBAP and current director general of IPGRI, Emile Frison; the current director of INIBAP, Richard Markham; Jamie Watts, Charles Staver, and Charlotte Lusty, impact specialists at IPGRI; Cliff Gold and other expert staff at IITA who facilitated and advised the research team; and Ruth Meinzen-Dick for her role in initiating and advising in the design of this research. We also thank an anonymous IFPRI reviewer and external reviewers for detailed and constructive comments on earlier drafts of this report.

Summary

This research report highlights findings from a set of studies undertaken by applied economists on the impact of improved banana cultivars and recommended management practices in the East African highlands. A particular focus of the analysis is genetic transformation of the cooking banana. Genetic transformation to achieve pest and disease resistance of the cooking banana is a promising strategy for smallholder farmers in this region. Biotic constraints are severe and not easily addressed through conventional breeding techniques or control methods. Exports on the world market are currently negligible, so that the risks of reduced exports due to policies against genetically modified foods are low. The crop is both an important food source and a significant generator of rural income, which means that improving productivity could have great social benefits.

Findings demonstrate the potentially pro-poor application of transgenic cooking varieties and the likely social consequences of the choice of host cultivar for trait insertion. Simulations illustrate the extent to which supporting public investments in education, market infrastructure, and extension would augment farmer demand for new planting material. Demand for planting material of potential host varieties for gene insertion varies according to household and physical farm characteristics, markets, and the attributes of varieties. In particular, farmers demand material with lower expected yield losses to black Sigatoka and weevils. The evidence also confirms that adoption of Fundación Hondureña de Investigación Agrícola (FHIA) hybrids reduces the vulnerability of Tanzanian households to yield losses from pests and disease. For farmers with few nonfarm sources of income, reducing production risk can smooth both consumption and income from local sales. In addition, it is evident that social capital plays a significant role in the use of recommended practices for managing soil fertility in banana production in Uganda. Most of the villages in the banana-producing areas foster active social organization, although membership in economic associations is more exclusive. Village social characteristics will likely have especially important implications for planting-material systems of East African highlands bananas. For this crop, transfers of planting material and related information are heavily farmer based, compared to the more formal seed distribution systems of some cereal crops.

Analysis of banana productivity confirms its dependence on the economic, climate, and soil characteristics of subregions. Evidence suggests that the efficiency of banana production can be improved, particularly in the Central Region of Uganda, where labor productivity is higher than elsewhere in the country. Soil pH and the application of manure have positive and significant effects on productivity, especially in southwest Uganda and the Central Region. The critical importance of labor in productivity is evident in findings regarding the age of the household head and family size. The analysis of production efficiency reveals the presence of surplus labor in banana production in the highlands, while access to farm labor is a constraint in the lowlands. Wages are higher for casual labor in the lowlands, where a more developed market for unskilled labor exists. Investment in technology development could shift banana productivity in the highlands. In the lowlands, both technology development and extension education have positive effects on productivity. Investing in human capital—especially women's education—–and enabling more access to input and credit markets could improve banana production, and investments in the paved road network would improve the comparative advantage of the lowlands in banana production.

Simulations of the gross economic benefits that could be generated by a set of technology options indicate that more widespread adoption of current best practices is likely to generate the greatest investment payoffs in Uganda. Yet the authors argue that the longer-term strategy endorsed by the National Agricultural Research Organization of Uganda, which combines conventional and transgenic approaches to mitigate the biotic pressures that cause major economic losses, is essential for sustaining banana-production systems.

The results of this study have implications for research and development (R&D) policies according to five categories:

1. *Developing Improved Banana Genotypes.* Findings confirm the vulnerability of the cooking banana, as well as other types grown in the region, to pests/diseases and the need for a research policy to commit long-term investments in the development of resistant genotypes. The choice of host cultivar and use group will have social consequences in terms of which farmers will be most likely to use and benefit from improved cultivars. In general, however, greater social costs are likely to be caused by delays in banana improvement than by choice of any host cultivar or technology set. Time lags in research and adoption have often been shown to be the single-most important determinant of the social payoff to investment.
2. *Enhancing Demand and Supply of Improved Germplasm.* A policy supporting investments in agricultural education, extension, marketing infrastructure, and access to good roads will enhance demand and supply of improved banana cultivars and, in turn, raise banana productivity and efficiency. We predict a demand for pest- and disease-resistant material, given the evidence that farmers value these traits, but demand will be much greater if other supporting investments in education, extension, and market infrastructure are made.
3. *Designing Effective Dissemination Mechanisms.* Findings support the current policy of the Government of Uganda, which emphasizes farmer association and human capital development as pillars of technology and knowledge dissemination. Farmers are price responsive, adopting new technologies at greater rates when output prices are high relative to input prices. The high rates of dissemination of improved practices for managing soil fertility are promising, despite the labor these practices demand. Farmer- and socially based mechanisms appear to be a crucial factor in the dissemination of both planting material and technologies. This finding reflects, in part, the clonally propagated nature of the banana plant.
4. *Scaling Up Genotype Adoption.* A demand-driven strategy for scaling up farmer use of approved banana varieties is needed. Widespread adoption of FHIA hybrids occurs in Tanzania. The analysis suggests that higher adoption rates in Tanzania relative to Uganda result from greater disease pressures, heavy dissemination efforts, and the fact that historically farmers have actively sought out planting material that is free of pests and diseases. Adoption definitely shows an impact on vulnerability to disease losses. Further analysis is needed over time to determine whether diffusion and benefits have been sustained, and whether incomes have been affected. The examples of farmer-to-farmer exchanges described in the report for Uganda, though limited in their impact in terms of numbers of farmer and communities, warrant closer examination as models for more structured and decentralized diffusion mechanisms. We recommend a farmer- and socially based network design, with farmer-supplied planting material, possibly scaling

up from some of the experiences in Uganda. The strategy of providing materials free-of-charge in large quantities is not sustainable.

5. *Developing R&D Strategies for the Highlands and Lowlands.* In the high-elevation areas, developing and promoting best cultural practices and marketing improvements are of primary importance. In the low-elevation areas, developing and promoting pest- and disease-resistant endemic cultivars is a priority, alongside cultural practices focusing on reviving productivity. To support the success of these efforts, major investments will need to be made in dissemination.

Technical change is a continuous, multidimensional process. Social science research can support the decisionmaking of research programs and investors by identifying the impediments that must be addressed to ensure that promising new technologies are in fact widely adopted by farmers. Genetic transformation of the cooking banana offers the rare opportunity to maintain the end-use qualities preferred by East African consumers while enhancing agronomic traits. Given the economic role of the banana plant in the region, the social benefits of technical change will be significant.

Acronyms and Abbreviations

AGT	Agro-Genetic Technologies, Limited, Kampala, Uganda
ARDI	Agricultural Research and Development Institute, Maruku, Tanzania
ARI	Agricultural Research Institute, Maruku, Tanzania
CEC	cation exchange capacity
CGE	computable general equilibrium
CIMMYT	International Maize and Wheat Improvement Center, Mexico City
CTA	Technical Centre for Agricultural and Rural Cooperation, Wageningen, The Netherlands
EAHB	East African highlands banana
FHIA	Fundación Hondureña de Investigación Agrícola, Tegucigalpa, Honduras
FOB	free on board
IFPRI	International Food Policy Research Institute, Washington, D.C.
IITA	International Institute for Tropical Agriculture, Kampala, Uganda
INIBAP	International Network for Improvement of Banana and Plantains, Montpellier, France
IPGRI	International Plant Genetic Resources Institute, Rome
ISAAA	International Service for the Acquisition of Agri-Biotech Applications
KARI	Kawanda Agricultural Research Institute, Kampala, Uganda
KCDP	Kagera Community Development Programme
KCDP project	five-year project titled Propagation and Diffusion of Superior Banana Cultivars, launched by KCDP in collaboration with the Belgian Technical Cooperation and Tanzanian authorities
LC3	local council level 3
MARIK	Makerere University Agricultural Research Institute, Kabanyolo, Uganda
m.a.s.l.	meters above sea level
ML	maximum likelihood
NARO	National Agricultural Research Organization, Kampala, Uganda
NBRP	National Banana Research Programme, Kampala, Uganda

NGO	nongovernmental organization
PSU	primary sampling unit
R&D	research and development
SOM	soil organic matter
SSU	secondary sampling unit
SWC	soil and water conservation
TE	technical efficiency
USAID	United States Agency for International Development, Washington, D.C.

Part I. Research Methods

CHAPTER 1

Assessing the Impact of Technical Innovations in African Agriculture

Melinda Smale

This chapter explains the motivation for the research reported in this monograph, placing it in the context of previous research about the impacts of technical innovations in African agriculture. While recognizing the limitations of impact assessment methods, it is argued that social science research can play a useful role in support of technology development. Interest in genetic transformation of East African highland bananas (EAHBs), among other options for enhancing banana production, is explained. Genetic transformation of the EAHB for pest and disease resistance represents a promising strategy for smallholder farmers, because (1) biotic constraints are severe and are not easily addressed through conventional breeding or methods of control; (2) exports on the world market are currently negligible; and (3) the crop is important for food as well as cash, generating rural income. Finally, contributions of this research are mentioned and the structure of the report is summarized.

Genetic improvement in seed (planting material, in the broadest sense) is an important component of agricultural growth for nations and regions in the process of economic development. Asia's Green Revolutions in rice and wheat are perhaps the most well-known examples of dramatic agricultural growth associated with crop genetic improvement. The process emphasized irrigated agriculture but was broad based, freeing labor and capital for the nonfarm economy and benefiting consumers through lower food prices (Lipton and Longhurst 1989; Rosegrant and Hazell 2000). Most experts agree that no such seed-based revolution has occurred in Sub-Saharan Africa, despite frequent, though scattered, episodes of success. Technical change in agriculture remains crucial to food supply and income generation in rural areas, however.

The research compiled here was motivated by the need, expressed both by those who conduct research on it and those who invest in it, to assess the economic impact of improved banana technology on smallholder farmers in the Lake Victoria regions of Uganda and Tanzania. Some of the improved technology explored in this report already exists, and some is emerging, including genetic enhancement through crossing, genetic enhancement through gene insertion to protect yields against pests and diseases, and soil fertility enhancement in banana production through current recommended practices.

Assessing Economic Impacts Ex Ante

The reliability of impacts assessed ex ante depends on the extent to which the past can be used to predict the future. Findings derived from past assessments of improved seed and related practices in Africa are inconclusive in two respects. Not only is their ability to encapsulate what happened in the relevant time period questionable, but their relevance to emerging gene-based technologies in today's societies may also be limited.

Past assessments have documented numerous examples of successful innovations in African agriculture (Dommen 1988; Gilbert et al. 1993; Sanders, Shapiro, and Ramaswamy 1996; Byerlee and Eicher 1997; Haggblade 2004), but often their impacts have been short-lived or difficult to substantiate in quantitative terms at a national or regional level of analysis. Once a technology has diffused among farmers and farming systems, disengaging the effects of genetic change from related inputs and environmental factors is a well-known challenge (Alston, Norton, and Pardey 1995; Morris and Heisey 2003). For improvements in pest and disease resistance, establishing the scenario that represents the absence of genetic change is especially challenging when it requires the estimation of farmers' yields over wide expanses in the presence of losses that are not observable. Experimental data on crop-yield losses reveals the severity of the disease or pest when it occurs but not its incidence across geographical regions and farm-management regimes. Most often, despite questionable assumptions, expert opinions are combined with experimental data in a sensitivity analysis that generates a range of estimates of the percentage of crop yields that would have been lost had the improved seed technology not been adopted (Marasas, Smale, and Singh 2003).

Broader social, political, and economic contexts mediate the impacts of seed technological change—which evolve over time but may also change abruptly. For example, the state-based public research and delivery systems that supported the diffusion of modern technologies during the decades following independence of many African nations subsequently proved to be fiscally unsustainable. Once dismantled, some state-based systems have been partially replaced by a patchy mosaic of firms and nongovernmental organizations. In other cases, former "control" policies have continued in a more contracted way, excluding smallholder farmers, while continuing to benefit politically important farming interests (Jayne et al. 2002).

Over the same period, the relative importance of agricultural income in the income of poor households has declined, so that the effects of seed technological change on employment and poverty may not be of the magnitude they once were (Meinzen-Dick et al. 2007). In Africa, market liberalization has progressed unevenly; farm families often cope with high and variable input as well as output prices, striving to meet their cash needs through numerous sources in an increasingly monetized economy (Bryceson 2002).

The failure of market liberalization has been attributed to the failure of policies to stimulate price signals to producers or to farm-level constraints that effectively block a supply response (Carter 2000). Encouraged by donors, governments have sometimes pursued a crisis-to-crisis approach to agricultural policy in food crops, leading to vacillating input–output price ratios for smallholders and inconsistent technology recommendations (for example, Heisey and Smale 1995, revisited a decade later by Smale and Jayne 2003). Policies that contributed to apparent success in one period later led to decline in smallholder use of the technology.

In the meantime, the number of institutional actors has expanded, and the configuration of institutions involved in agricultural R&D, as well as technology provision to farmers, has changed. The productivity gains of the Green Revolution occurred in a narrow range of crops and technologies, financed by commodity-oriented, supply-driven, public national research programs backed by international agricultural centers (Byerlee and Echeverría 2002). "Orphan" crops, such as the cooking banana, have fallen largely outside the mandates of the international system, but at the same time they have not proved profitable to private investors (Naylor et al. 2004).

Roles are evolving, however, and public–private dichotomies are less useful in analyzing institutional change. A variety of institutional modes will be necessary to expand the range of technologies and more fully address issues related to their social, economic, and environmental consequences. These include private funding through producer associations (for commercial, tradable crops) and various forms of public–private partnerships, alongside continued public funding for essential public goods, such as research on managing natural resources and the environment, prebreeding, and conserving germplasm. The need to work through civil society and other institutions to enable the successful integration of technological innovations in African agriculture has become increasingly evident (Haggblade 2004).

Despite these changes and other calls for change, the paucity of public funds in Africa—and its implications—cannot be understated. Africa received only 8 percent of public agricultural research expenditures in 1976 and 6 percent in 1995, with the lowest overall rate of increase among all regions for the two decades, and it is the only region of the world where per capita research spending, in terms of both total population and agricultural workers, declined (Pardey and Beintema 2001).

Role of Social Science Research in Technology Development

Social science research can help identify ways to create an enabling environment for improved crop cultivars and practices before their release. There is a need to diagnose impediments and constraints to adoption of cultivars during the development of the cultivar in order to contribute to the future success of the cultivar in farmers' fields. Some challenges associated with crop biotechnologies are the same as those faced with conventional genetic technologies.[1] This commonality is especially true with respect to the decisions of the farmers, who are the potential clients for the research product.

For example, effective, demand-driven provision of planting material is a critical ingredient to any successful seed innovation. Poorly developed markets for planting material, weak institutions for diffusing it, or the extreme poverty and cash-flow problems experienced by many smallholder farmers in Sub-Saharan Africa have often thwarted farmers' ability to benefit from improved cultivars, even when these cultivars perform well in their fields (Tripp 2003). The agricultural R&D institutions that constitute the commercial seed system in high-income countries deliver final products to dispersed clients, who demand them on a regular basis. Compared with these well-served farmers, those in marginal environments of poorer countries in Sub-Saharan Africa often rely on their own saved seed or village connections, because seed supplied through more formal market channels is unreliable. Planting-material systems for clonally propagated crops, such as banana, are characterized by a far greater degree of farmer-to-farmer exchange, and some public investment is likely to be required to support the dissemination of improved bananas, whether or not product marketing channels are well developed (de Vries and Toenniessen 2001).

Another challenge common to any new technology is that farmers must perceive its benefits in their fields, not just on the ex-

[1] Here, "conventional genetic technologies" refers to any type of hybrid (for example, varietal, top-cross, three-way, and single- or double-cross), improved open-pollinated cultivars (such as a synthetic or composite) or pure line selection developed in a crossing or selection program, in contrast to those developed through gene insertion.

periment station. Not all new seed cultivars are widely grown, or "popular"—especially in semicommercial, smallholder production in Africa, where farmers have multiple objectives and face combinations of biophysical and economic constraints. Farmers may not discern the benefits from inserting the trait, or may view these as less important than some other disadvantageous traits of the new cultivar relative to currently grown cultivars. Even if planting material is available, and they can afford to purchase it, farmers will not adopt a cultivar unless the benefits are perceptible. For example, resistance to diseases and pests are some of the traits targeted with gene-based technologies. Evidence suggests that although farmers are knowledgeable about pests that they can see and touch (weeds, insect, vertebrate pests), they know much less about plant diseases and insect reproduction (Bentley 1994; Orr 2003), which scientists observe with the assistance of laboratories and controlled experiments. Incidence varies over years and within crop stands, following a statistical distribution that farmers observe incompletely.

As with any introduced technology, knowing the social determinants and social consequences of adoption is important for designing policy interventions to support it. Though planting material may be neutral to the scale of the farm operation (meaning that there is nothing inherent in the technology that implies large-scale farmers will have greater ability to use it than will smallholder farmers), there is typically an aspect of the technology that favors its adoption by certain social groups. Reviews of related literature for other periods and regions confirm that whether a technology benefits poor people depends more on underlying social and economic conditions than on the intrinsic attributes of the technology (Feder, Just, and Zilberman 1985; Feder and Umali 1993; Hazell and Haddad 2001; Meinzen-Dick et al. 2007).

Compared with conventional technologies, some challenges are unique to genetically transformed seed, such as the need to develop appropriate biosafety regulatory frameworks (Cohen and Paarlberg 2004). Governments have a public responsibility to invest in assisting farmers to make informed choices and to take full advantage of the technology if they choose to use it (Tripp 2001; Smale and de Groote 2003). Citizens of African nations will want to make their own informed decisions.

Crop Biotechnology for African Smallholders

Technological innovations are undoubtedly necessary for economic change and growth, and there is no question that feasible technology options can be found on the research shelves of African nations. A recent expert survey revealed 40 biotechnology products in the public research pipeline for Kenya, South Africa, and Zimbabwe alone (Atanassov et al. 2004; Cohen 2005). Only in South Africa, however, have transgenic crop cultivars been released to farmers.

Given this situation, assessment of the technologies is by definition ex ante, or based on prediction. The case study summarized in this report was selected from among products now in research pipelines according to several criteria. It can be argued that the biotechnology innovations with greatest promise for smallholder farmers in Africa today are those that (1) tackle economically important biotic or abiotic constraints that are not easily addressed through conventional plant breeding or methods of control (de Vries and Toenniessen 2001); (2) pose little risk of endangering trade through exports to countries that do not accept transgenic products (Nielsen, Thierfelder, and Robinson 2001); and (3) can make a difference in the welfare of smallholder farmers by serving as either sources of food or cash.

First, to be cost effective, biotechnology tools should demonstrate a comparative advantage relative to other tools or tool combinations. To target traits effectively and result in popular crop varieties, genes must be inserted into well-adapted host cultivars that are either conventionally bred or farmer selected. The second criterion acknowledges that genetic engineering of important export crops makes them vulnerable to trade disputes, regulations, and political lobbies outside their borders. For traded commodities, full market liberalization implies that shifts in agricultural productivity may not translate into lower food prices. For nontradable food crops, downward pressure is exerted on prices by seed technological change because they are determined within borders rather than on the international market.

This observation relates to the third criterion, because the vast majority of smallholder farmers in Sub-Saharan Africa consume part of their food production and many are net consumers. Both urban and rural consumers in these countries will benefit many times more from the price decreases that accompany technological change than will those of richer countries, because food occupies a large proportion of the budget of low-income people, and they are more responsive to price changes in terms of quantities demanded (Pinstrup-Andersen and Cohen 2001).

Reduction of crop losses and yield increases can still lead to a significant increase in the income of poor households if markets function well. Focusing on the productivity of food crops, when bolstered by stronger markets, will also have national and regional income effects. Recent analyses confirm the potential role of food staples in domestic and intraregional markets as a source of growth in demand for African agricultural products and income, in comparison with traditional and nontraditional export crops (Diao and Hazell 2004).

In East Africa, the case of bananas genetically transformed for resistance to pests and disease meet these criteria. East Africa is the largest banana-producing and banana-consuming region in Africa. This research focuses on the principal banana-growing areas around Lake Victoria in Uganda and Tanzania, which form a contiguous area stretching from the northeast portion of the lake, through the historical locus of Ugandan banana production in the Central Region (including the new locus with more market-oriented production in the southwest highlands of Uganda), to the Kagera Region in northwest Tanzania. The East African highlands banana (EAHB) is the dominant type grown throughout this expanse, and a set of common biophysical factors constrain productivity. As the chapters of this report show, bananas are produced largely for food consumption, with surpluses sold in town and city markets. Growing urban markets contribute to the importance of the crop as a source of cash. Small quantities enter regional trade, with negligible exports. The population of the area is related linguistically and culturally, though separated by borders and administrative divisions, as well as by important political and economic differences.

Though the origin and center of diversity for bananas is believed to be Southeast Asia (Simmonds 1959), the East African highlands is recognized as a second center of diversity for an endemic genomic group (AAA-EA), which comprises two groups defined by their use in cooking and brewing local beer. The endemic group is also classified in terms of clone sets. All EAHBs are triploids. With three genomes, as opposed to two or four, few cultivars are fertile. Though professional breeding of bananas began during the 1920s, it was not until recently that a major breakthrough was achieved through the development of a hybridization technique by the FHIA. Fertile male diploids are used to pollinate triploid cultivars to produce tetraploid hybrids. These are crossed again with improved diploids to give secondary triploids, from which selections for release to farmers are obtained. However, the endemic cultivars most preferred by farmers are infertile and cannot be improved by conventional means. This difficulty of improving bananas by means of conventional plant breeding techniques (based on seed production by either self-pollination, cross-pollination, or by hybridization) makes genetic engineering a particularly attractive means of combating pests and disease through host-plant resistance (see Chapter 4 and Johanson and Ives 2001). In addition, even when breeders choose parents preferred by farmers, the development of hybrids through conventional crosses often produces bananas that are distant in type from either parent. Thus, biotechnology offers a means of genetic improvement without transforming the end products that have proven to be of economic value to farmers and consumers. So far, genetic transformation for fungal resistance has been initiated in Cavendish and plantain types only but has been recommended for endemic types (Quemada and Johanson 2004). Farmers reproduce banana cultivars through vegetative propagation.

Contribution of This Research

This study contributes information in several ways. First, it provides a case study of the potential impacts of a food-crop biotechnology in smallholder African agriculture. From case studies published about the adoption of crop biotechnologies so far, relatively little can be gleaned about the potential demand for transgenic cultivars of staple food crops in Sub-Saharan Africa. Several ex ante studies predicting the economic impact of transgenic food crops have been conducted in the economic surplus framework (Qaim 1999a,b), based on assumed, rather than actual, adoption parameters. Farmer adoption of genetically improved cultivars has been analyzed ex post in high-income countries with fully commercialized agricultural production, based on field trials, farmer and consultant surveys, or field-level surveys (reviewed in Fernandez-Conejo and McBride 2002; Marra, Pardey, and Alston 2003).

Few biotechnology products have been released to farmers in agricultural economies that are not heavily industrialized. Recent research has analyzed the adoption and welfare impacts ex post of introducing transgenic cotton in China, South Africa, Mexico, Argentina, and India (see review by Smale, Zambrano, and Cartel 2006). Cotton, the championed success story in developing countries, is a special case. Cottonseed cannot be easily saved by smallholder farmers, and the requirements of ginning and delinting processes favor the vertical integration of cotton production and marketing. The story of Makhathini Flats of Kwa Zulu, Natal, illustrates the vulnerability of growers to such market arrangements (Gouse et al. 2005).

A second contribution of this research is the representative statistical baseline that it provides to national research programs for characterizing banana growers and bananas today and measuring the effects of projects in the future. A third is the integration of several in-depth studies, undertaken principally as doctoral theses (Edmeades 2003; Katungi 2006; Bagamba 2007; Nkuba 2007), which investigate specific components of banana production and types of banana cultivars and management practices. The principal objectives of these theses were to analyze

1. constraints to banana productivity and efficiency, including labor markets and soil fertility;
2. banana cultivar choice and demand with a trait-based approach, accounting for farmers' perceptions of biotic constraints and cooking quality;
3. the role of social capital and social networks in diffusion of best practices for soil fertility management in banana production; and
4. the effects of banana hybrids on smallholder farmers in the Kagera Region of Tanzania.

Drawing on the findings of these theses permitted a more comprehensive understanding of farmers' conditions. Viewing technology as a gradation of practices and innovations has the disadvantage of complexity but the advantage of providing insights that might not be possible when specific technologies are analyzed piecemeal.

Structure of the Report

The chapter that follows presents an overview of the study's conceptual approach, including the sample survey design. Given the need to address both a new type of genetic technology and a less researched crop, the collection and analysis of primary data and the adaptation of analytical methods played a relatively large part in this research. In Part II, features of the banana industry in Uganda and Tanzania and the historical contexts for developing genetic and other technical innovations in banana production are summarized. The baseline data are used to characterize banana growers and the bananas they grow. The chapters in Part III represent the various components of economic analysis undertaken by Ph.D. students; the National Agricultural Research Organization (NARO), Uganda; the Agricultural Research and Development Institute (ARDI), Maruku, Tanzania; and researchers from the International Food Policy Research Institute (IFPRI) to address specific research questions about economic impacts of improved cultivars and best management practices in the context of overall banana productivity and production constraints. Additional details related to these chapters can be found in the relevant theses. Part IV includes a summary of findings and implications for research and policy. The appendixes A–E include complete banana taxonomies for the two countries in the survey domain, village-level rates of adoption and village social structure in Uganda, and details of the sample survey design.

References

Alston, J. M., G. W. Norton, and P. G. Pardey. 1995. *Science under scarcity: Principles and practice for agricultural research evaluation and priority setting.* Wallingford, U.K.: CAB International.

Atanassov, A., A. Bahieldin, J. Brink, M. Burachik, J. I. Cohen, V. Dhawan, R. V. Ebora, J. Falck-Zepeda, L. Herrera-Estrella, J. Komen, F. C. Low, E. Omaliko, B. Odhiambo, H. Quemada, Y. Peng, M. J. Sampaio, I. Sithole-Niang, A. Sittenfeld, M. Smale, K. R. Sutrisno, R. Valyasevi, Y. Zafar, and P. Zambrano. 2004. To reach the poor. Results from the ISNAR-IFPRI *Next Harvest* study on genetically modified crops, public research, and policy implications. Environment and Production Technology Division Discussion Paper 116. Washington, D.C.: International Food Policy Research Institute.

Bagamba, F. 2007. Market access and agricultural production: The case of banana production in Uganda. Ph.D. thesis, Wageningen University, the Netherlands.

Bentley, J. W. 1994. Facts, fantasies and failures of farmer participatory research. *Agriculture and Human Values* 11 (2–3): 140–150.

Bryceson, D. F. 2002. The scramble in Africa: Reorienting rural livelihoods. *World Development* 30 (5): 725–762.

Byerlee, D., and R. G. Echeverría. 2002. *Agricultural research in an era of privatization.* Wallingford, U.K.: CAB International.

Byerlee, D., and C. K. Eicher. 1997. *Africa's emerging maize revolution.* Boulder, Colo.: Lynne Rienner.

Carter, M. 2000. *Land ownership inequality and the income distribution consequences of economic growth.* Working Paper 201. Helsinki: World Institute for Development Economics Research.

Cohen, J. I. 2005. Poorer nations turn to publicly developed GM crops. *Nature Biotechnology* (1): 27–33.

Cohen, J. I., and R. Paarlberg. 2004. Unlocking crop biotechnology in developing countries—A report from the field. *World Development* 32 (9): 1563–1577.

de Vries, J., and G. Toenniessen. 2001. *Securing the harvest: Biotechnology, breeding and seed systems for African crops.* Wallingford, U.K.: CAB International.

Diao, X., and P. Hazell. 2004. *Exploring market opportunities for African smallholders.* 2020 African Conference Brief 6. Washington, D.C.: International Food Policy Research Institute.

Dommen, A. 1988. *Innovation in African agriculture.* Boulder, Colo.: Westview Press.

Edmeades, S. 2003. Variety choice and attribute trade-offs within the framework of agricultural household models: The case of bananas in Uganda. Ph.D. thesis, North Carolina State University, Raleigh.

Feder, G., and D. Umali. 1993. The adoption of agricultural innovations: A review. *Technological Forecasting and Social Change* 43: 215–239.

Feder, G., R. Just, and D. Zilberman. 1985. Adoption of agricultural innovations in developing countries: A survey. *Economic Development and Cultural Change* 33(2): 255–296.

Fernandez-Cornejo, J., and W. McBride. 2002. *Adoption of bioengineered crops.* ERS Agricultural Economics Report AER810. Washington, D.C.: U.S. Department of Agriculture.

Gilbert, E., L. C. Phillips, W. Roberts, M. T. Sarch, M. Smale, and A. Stroud. 1993. *Maize research impact in Africa: The obscured revolution.* Washington, D.C.: USAID, Division of Food, Agriculture and Resources Analysis, Office of Analysis, Research and Technical Support, Bureau for Africa.

Gouse, M., J. Kirsten, B. Shankar, and C. Thirtle. 2005. Bt cotton in KwaZulu Natal: Technological triumph but institutional failure. *AgBiotechNet* 7 (ABN 134) : 1–7.

Haggblade, S. 2004. *Building on successes in African agriculture.* 2020 Vision for Food, Agriculture, and the Environment. Focus 12. Washington, D.C.: International Food Policy Research Institute.

Hazell, P., and L. Haddad. 2001. *Agricultural research and poverty reduction: Food, agriculture and the environment.* Discussion Paper 34. Washington, D.C.: International Food Policy Research Institute.

Heisey, P. W., and M. Smale. 1995. *Maize technology in Malawi: A Green Revolution in the making?* CIMMYT Research Report 4. Mexico, D.F.: International Maize and Wheat Improvement Center.

Jayne, T. S., J. Govereh, A. Mwanaumo, J. Nyoro, and A. Chapoto. 2002. False promise or false premise? The experience of food and input market reform in eastern and southern Africa. *World Development* 30 (11): 1967–1985.

Johanson, A., and C. Ives. 2001. An inventory of agricultural biotechnology for the Eastern and Central Africa region. Michigan State University, Institute of International Agriculture, Ann Arbor. Manuscript.

Katungi, E. 2006. Social capital and technology adoption on small farms: A case of banana production technology in Uganda. Ph.D. thesis, University of Pretoria, Pretoria, Republic of South Africa.

Lipton, M., with R. Longhurst. 1989. *New seeds and poor people.* Baltimore: Johns Hopkins University Press.

Marasas, C. N., M. Smale, and R. P. Singh. 2003. The impact of productivity maintenance research: The case of leaf rust resistance breeding in CIMMYT-related spring bread wheat. *Agricultural Economics* 29 (3): 253–263.

Marra, M., P. Pardey, and J. Alston. 2003. The payoffs to transgenic field crops: An assessment of the evidence. *AgBioForum* 5 (2): 43–50.

Meinzen-Dick, R., M. Adato, P. Hazell, and L. Haddad. 2007. Findings of an integrated economic and social analysis. In *Agricultural research, livelihoods, and poverty: Studies of economic and social impacts in six countries,* ed. M. Adato and R. Meinzen-Dick. Baltimore: Johns Hopkins University Press.

Morris, M. L., and P. W. Heisey. 2003. Estimating the benefits of plant breeding research: Methodological issues and practical challenges. *Agricultural Economics* 29 (3): 241–252.

Naylor, R. L., W. P. Falcon, R. M. Goodman, M. M. Jahn, T. Sengooba, H. Teferu, and R. J. Nelson. 2004. Biotechnology in the developing world: A case for increased investments in orphan crops. *Food Policy* 29 (1): 15–44.

Nielsen, C. P., K. Thierfelder, and S. Robinson. 2001. *Genetically modified foods, trade, and developing countries.* Trade and Macroeconomics Division Discussion Paper 77. Washington, D.C.: International Food Policy Research Institute.

Nkuba, J. M. 2007. Assessing adoption and economic impacts of new banana varieties on livelihoods of farmers, Kagera Region, Tanzania. Ph.D. thesis, Sokoine University, Sokoine, Tanzania.

Orr, A. 2003. Integrated pest management for resource-poor African farmers: Is the emperor naked? *World Development* 31 (5): 831–845.

Pardey, P. G., and N. M. Beintema. 2001. *Slow magic: Agricultural R&D a century after Mendel.* Agricultural Science and Technology Indicators Initiative. Washington, D.C.: International Food Policy Research Institute.

Pinstrup-Andersen, P., and M. J. Cohen. 2001. Rich and poor country perspectives on biotechnology. In *The future of food: Biotechnology markets and policies in an international setting,* ed. P. G. Pardey. Washington, D.C.: International Food Policy Research Institute.

Qaim, M. 1999a. *Assessing the impact of banana biotechnology in Kenya.* ISAAA Brief 10-1999. Ithaca, N.Y.: International Service for the Acquisition of Agri-Biotech Applications.

———. 1999b. Potential benefits of agricultural biotechnology: An example from the Mexican potato sector. *Review of Agricultural Economics* 21 (2): 390–408.

Quemada, H., and A. Johanson. 2004. Transgenic approaches to solving major pest and disease constraints in banana and plantain: An assessment of current status and recommendations for future information needs. Michigan State University, Institute of International Agriculture, Ann Arbor. Manuscript.

Rosegrant, M., and P. Hazell. 2000. *Transforming the rural Asian economy: The unfinished revolution.* Oxford: Oxford University Press.

Sanders, J., B. I. Shapiro, and S. Ramaswamy, 1996. *The economics of agricultural technology adoption in semiarid Sub-Saharan Africa.* Baltimore: Johns Hopkins University Press.

Simmonds, N. W. 1959. *Bananas.* London: Longman's.

Smale, M., and H. de Groote. 2003. Diagnostic research to enable adoption of transgenic crop cultivars by smallholder farmers in Sub-Saharan Africa. *African Journal of Biotechnology* 2: 586–595.

Smale, M., and T. S. Jayne. 2003. *Maize in eastern and southern Africa: "Seeds" of success in retrospect.* Environment and Production Technology Division Discussion Paper 97. Washington, D.C.: International Food Policy Research Institute.

Smale, M., P. Zambrano, and M. Cartel. 2006. Bales and balance: A review of the methods used to assess the economic impact of Bt cotton on farmers in developing economies. *AgBioforum* 9(3): 195–212.

Tripp, R. 2001. Can biotechnology reach the poor? The adequacy of information and seed delivery. *Food Policy* 26 (3): 249–264.

———. 2003. *Strengthening the enabling environment for agricultural technology development in Sub-Saharan Africa.* ODI Working Paper 212. London: Overseas Development Institute.

CHAPTER 2

Elements of the Conceptual Framework and Sample Survey Design

Melinda Smale, Svetlana Edmeades, Stanley Wood, and Robert Kalyebara

This chapter presents specific questions posed in this research, elements of the conceptual framework, and a synopsis of the sample design. Two levels of observation and analysis are used. At the farm level, the underlying framework is the model of the agricultural household. The economic surplus approach is applied at the industry level. Details of the models are given in each chapter of Part III. The sample villages and households are stratified by elevation and exposure of the surrounding locality to introduction of banana planting material.

Research Questions

Diagnostic research investigates causal relationships and describes the nature of a situation. The opportunity to conduct extensive diagnostic research that is farm based during the scientific process of crop improvement is less common than perhaps it should be. Three general questions were posed. First, what is the status of existing banana cultivars and management practices, as well as constraints to production and marketing, on farms in the Lake Victoria region of Uganda and Tanzania? Second, given these and other regulatory constraints, what are the prospects that banana growers will adopt cultivars with transgenic resistance to pests and diseases? Third, what is the potential impact of emerging banana technologies, including transgenic cultivars, on the banana industry, assuming that farmers adopt them? To answer these questions, analysis was conducted at the farm and industry levels, with a combination of tools.

Levels of Observation and Analysis

Each level of observation and analysis contributes information, and each has limitations. Detailed farm-level analysis from a statistically representative sample is used to identify the determinants of technology use, predict demand for planting material, and test hypotheses about production efficiency and labor use. The characterization of the banana growers and banana cultivars furnishes a baseline.

Findings from farmer surveys can be generalized only to the extent that the samples are representative of the full geographical domain over which there are potential economic payoffs from technology use. Analyses of farmer adoption also fail to provide information about the economywide impacts of technical change or the social distribution of benefits at the national and regional levels. Industry or mesoscale analysis, conducted across larger geographic scales and using secondary or more aggregated forms of primary data, can be used to examine some aspects of technical change that farmer surveys alone cannot address. For example, investment in R&D in Uganda could affect geographic areas not intentionally targeted, such as the Kagera Region of Tanzania. From a producer perspective, the impact of widespread technology use can be negative if it exerts downward pressure on prices, because crop supply grows faster than crop demand.

An appropriate tool to answer questions concerning the expected total magnitude and distribution of economic benefits from the diffusion of various technologies at the national and regional levels is an economic surplus model, such as that documented fully in Alston, Norton, and Pardey (1995) and applied in Chapter 6. In general, this type of analysis involves the simulation of benefit outcomes based on carefully constructed scenarios. Scenarios might include different technologies or price or trade policies. Scenarios are then compared according to investment criteria. Criteria often include the size of social and private benefits relative to health or environmental risks and benefit shares earned by consumers, commercial or smallholder producers, or regions.

However, the economic surplus approach also has known limitations. First, the approach is sector based, representing partial changes in market equilibrium. Economywide effects can be traced through the application of a multimarket model or a computable general equilibrium (CGE) model. These methods are not as informative about the incidence of benefits and costs across economic agents and social groups in the industry as the economic surplus approach, however.

A drawback that is common to the economic surplus approach and multimarket and CGE models is that they gloss over a number of intervening factors and conditions that must be met for the release of the technology to actually result in a shift of the industry supply curve. Ex ante models based on the economic surplus approach depict supply-driven effects, but take adoption parameters as given. For this reason, in Chapter 6, we also estimate the adoption parameter and its determinants, and draw inferences about the social consequences of inserting resistance into one host cultivar as compared to another. The appropriate analytical framework for the models developed and estimated at the farm level is the theory of the agricultural household, described next.

Technology Use on Farms

The performance of a technology (crop and trait), either on experimental stations or on farms, is only one consideration among many in its adoption. Once a technology has been developed and tested, factors that have incidence at national, regional, and local levels influence whether smallholder farmers will choose to use it, the geographical extent of use, and its continuity and duration. For example, biophysical characteristics of an agroecological region are important determinants of use. The economic impact or "success" of the technology is determined in the first instance by use.

We often employ the term "use" rather than "adoption" in recognition of the complexity of defining adoption. Adoption can be represented by a discrete decision to use or not to use (a variable defined as either one or zero), a proportional indicator (such as a share of land or share of plants), an index of scale or extent of use (the number of hectares or the number of plants), level choices or intensity of use (per hectare, or per plant), or frequency of use (number of seasons or number of applications per season). Use at any one point in time is not a robust indicator of "adoption." As mentioned in Chapter 1, there are many examples of adoption discontinuities caused by either technical or institutional factors.

Once a crop cultivar has been released, farmers' choices are influenced by market conditions for seed, related inputs, and the crop, as well as their surrounding agroecosystem (soils, pests and plant disease pressures, moisture, elevation, and environmental heterogeneity). Explanatory factors that vary among farmers are documented in a broad literature on adoption of agricultural innovations in developing countries (reviews include Feder, Just, and Zilberman 1985; Feder and Umali 1993). Informal organizations, village associations, and social networks may substitute or complement extension services as sources of the information that is critical for farmers to perceive and earn benefits from cultivars or practices.

The agricultural household model is used as the conceptual framework in the four chapters of this report analyzing the use and impact of banana technologies at the farm level (Chapters 6–9). Input markets, in particular, are often incomplete for banana growers in Uganda and Tanzania. In this framework, the agricultural household maximizes utility (u) from consumption of banana goods (x^{B}), other purchased goods (x^{G}), and leisure or home time (h), conditioned on a set of household characteristics ($\mathbf{\Omega}_{HH}$) over a vector of choice variables, defined as ψ:

$$\max_{\psi} u\left[x^{B}, x^{G}, h \mid \mathbf{\Omega}_{HH}\right].$$

The household maximizes utility subject to a number of constraints. The production technology is defined by a set of variable inputs, denoted by a vector $\mathbf{N}$, used for the production of banana output (q) on a predetermined amount of land (A) in the banana grove or plantation. Variable inputs comprise primarily planting material, labor, and organic fertilizers. The production technology is conditioned on the physical characteristics of the farm ($\mathbf{\Omega}_{F}$):

$$g\left[q, \mathbf{N}, A \mid \mathbf{\Omega}_{F}\right] = 0.$$

Household budget limitations are depicted by the full income constraint for all traded banana products (T), where $\mathbf{p}^{B}$ is a vector of banana product prices, p^{G} is the price of other goods consumed by the household, $\mathbf{p}^{N}$ is a vector of input prices (for example, the wage rate, w, being one element of this vector) and I is exogenous income:

$$\sum_{j \in T}(q_j - x_j^{B})\,\mathbf{p}_j^{B} - p^{G}x^{G} - \mathbf{p}^{N}\mathbf{N} + I = 0.$$

Semisubsistence households may face sufficiently large transactions costs of market participation that they opt for autarky with respect to some goods or services (de Janvry, Fafchamps, and Sadoulet 1991). For nontraded goods (NT), households equate quantities produced and consumed:

$$q_j - x_j^{B} = 0, j \in NT.$$

With imperfect input markets, constraints on factors of production are also imposed, depending on the context of the model, including time, planting material, or land constraints. This general analytical approach is then adapted by authors to test specific research hypotheses. The structure of household preferences, choice variables, and explanatory variables in the reduced-form equations depends on the application. Chapter 6 formulates a trait-based model of cultivar demand. Farmers' planting decisions are hypothesized to be a function of the consumption attributes and production traits of bananas, controlling for other household, farm, and market characteristics. To explain the use of recommended soil fertility practices in banana production, Chapter 7 introduces social capital in addition to these factors. Chapter 8 estimates the productivity and efficiency of banana production, focusing on soil characteristics and labor use. The adoption equation estimated in Chapter 9 builds on the same general framework to identify the determinants of hybrid use and test the impact of adoption on production vulnerability.

Establishing a Counterfactual

In the literature about assessing the effects of agricultural research, the "factual" describes the state or situation in the presence of technological change from the adoption and diffusion of new crop cultivars or crop management techniques; the "counterfactual" refers to the situation in the absence of the technology. Methodological and conceptual challenges have long plagued attempts to define the two states and separate the effects of the technology per se from countervailing social, political, and economic changes that occurred simultaneously and may have been driven by similar underlying factors (Kerr and Kolavalli 1999; Meinzen-Dick et al. 2007).

The controlled conditions achievable during the implementation of physical experiments are not an observable but a heuristic state. Even in a controlled biological experiment, two types of problems are involved in measuring the effects (typically a continuous variable) of a treatment (often a dichotomous variable). First, the effect of the treatment is heterogeneous, varying across individuals. Heckman developed both a general model (Heckman 1990) and a simple two-step method to address selection bias (Heckman 1976) that have been widely applied in the adoption literature. The proliferation of applications of this influential approach has since led to some concern for their quality (Johnston and Di Nardo 1997). For example, the parameters of the model appear be very sensitive to heteroskedasticity, and the approach is relatively inefficient compared to maximum likelihood estimation. Instead, simpler estimation approaches involving instrumental variables are often employed, though finding variables that are correlated with treatment but not outcome can be difficult. Recently, Duflo and Kremer (2003) have argued that there is considerable scope for greater use of randomized evaluation methods in addressing selection biases in impact evaluation. Though such methods eliminate selection bias by construction, they generate highly location-specific analyses of project interventions that may ignore the numerous factors operating at a larger geographical scale of analysis—such as agroecological differences, variation in market infrastructure, prices, and other institutional and policy factors. These are some of the most important factors that influence the capacity of individual households to respond to changes in the agricultural economy.

In addition to this generic dilemma faced by any social scientist assessing the impacts of improved crop cultivars, we have faced several particular challenges in this research. First, our "factual" is itself a prediction, because no transgenic cultivars of banana have yet been released. Second, the "counterfactual" is difficult to define because of the range of banana types currently grown by farmers. For example, farmers grow numerous clones of endemic highland bananas. They also grow "elite" or "superior" farmers' cultivars from the region that have been multiplied through tissue culture for dissemination by the national research program. Farmers also grow "exotic" cultivars, which are farmers' cultivars that have been introduced from other regions of the world. More recently, hybrids have been released and introduced into the region. That is, several of these counterfactuals themselves represent the ex post analysis of agricultural research, conducted here for the first time.

To retain the cooking quality that is preferred in Uganda, the National Agricultural Research Organization (NARO) has targeted cooking bananas for genetic transformation. FHIA hybrids are dessert bananas that have been used by farmers for multiple purposes. Consequently, some of the factors affecting adoption of FHIA hybrids will not be the same as those that will affect the transgenic bananas currently being developed. Chapter 6 examines the potential for transgenic cooking cultivars in Uganda, considering endemic cultivars and elite farmers' cultivars as the counterfactual. Chapter 9 investigates the adoption of FHIA hybrids in Tanzania, compared to endemic cultivars and farmers' exotic cultivars.

We addressed the problem of selection bias through a combination of sampling methodology and application of a treatment model (Chapter 9). The sample survey design is summarized next, with a more complete description in Appendix D.

Sample Survey Design

The population domain was selected to cover areas specializing in banana production, including those with decreasing, unchanging, and increasing levels of production. These correspond roughly to the central and southwest geographical zones in Uganda, and the Kagera Region of Tanzania.

A geo-referenced map of East Africa indicating principal banana growing areas was provided by the International Network for the Improvement of Banana and Plantains (INIBAP) to delineate the domain. Because there are trade-offs in precision as the number of stratifying variables increases with a fixed sample size, we confined the number to two.

The first is environmental. A critical parameter for the adoption of cultivars with improved resistance to pests and disease is the yield advantage attained relative to other banana cultivars. Other than farmers' management practices, which are part of farmer decisionmaking (and thus, cannot serve as a control variable), relative yield advantages depend on the biotic stresses and productivity potential of the growing environment. In consultation with INIBAP, International Institute of Tropical Agriculture (IITA), NARO, and ARDI scientists, elevation was selected to represent the numerous correlated factors that affect the incidence and severity of most pests and diseases of bananas in the Lake Victoria region (Speijer et al. 1994). Elevation is also related to soil quality, climate, and the surrounding vegetation in these environments (Tushemereirwe et al. 2001). Elevation was delineated at 1,200 m above sea level (m.a.s.l.), defining low elevation as below 1,200 m.a.s.l. and high elevation as above this value.

The second stratifying variable was previous exposure to new banana cultivars (exposed; not exposed). Areas of exposure were defined as a local council level 3 (LC3, or subcounty) or ward where researchers, extension agents, or other program agents had introduced improved planting material (banana suckers) in at least one community. Areas with no exposure were defined as those where no organized program designed to diffuse improved planting material had been conducted, according to personal consultation with NARO and ARDI, records of the Kagera Community Development Programme (KCDP), and other program records. Areas included in exposed strata represent the factual and those included in the nonexposed strata represent the counterfactual in predicting impacts of improved banana cultivars.

Our intention in defining "exposure" at a broader geographical scale than the village was to ensure that we accounted for the possible transmission of planting material and information among as well as within villages. Clearly, although we expect to have reduced the selection bias at the household level, we have not eliminated the bias associated with the decision to promote new planting material in one administrative area rather than another ("program placement").

Geo-referenced data about banana-production systems, a digital elevation model, maps of administrative units, and information concerning previous diffusion of banana planting material were used to disaggregate the domain into a total of four strata (i = elevation, j = exposure): (1) low elevation, with exposure ($i = 1, j = 1$); (2) low elevation, without exposure ($i = 1, j = 0$); (3)

Figure 2.1 Sample domain: Elevation overlaid with exposure/nonexposure

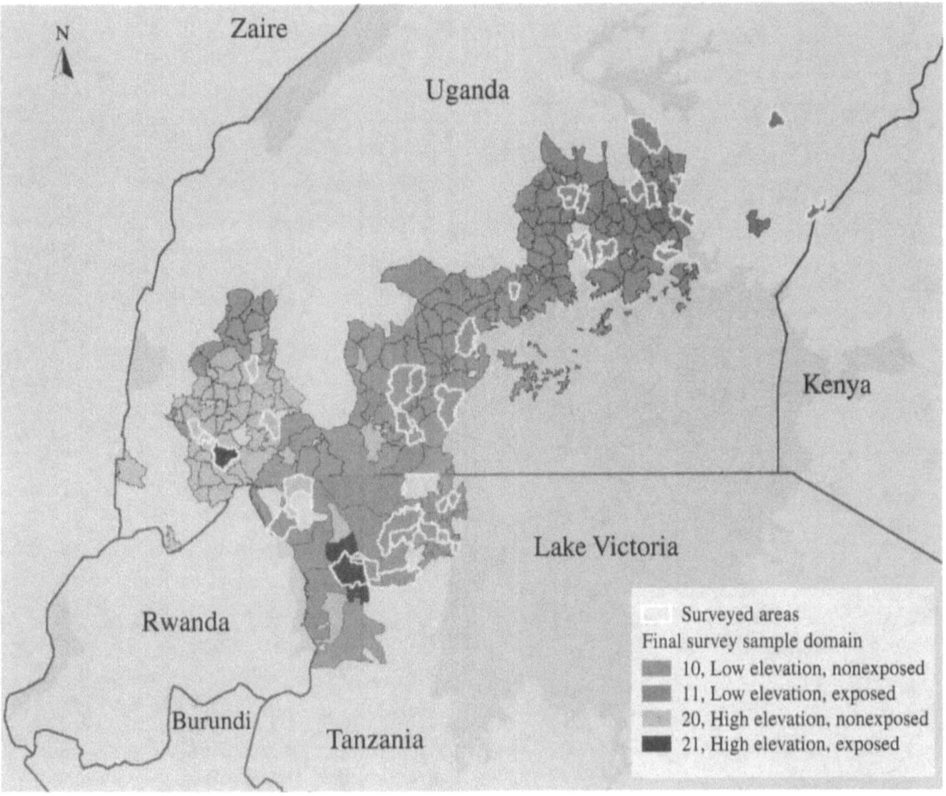

high elevation, with exposure ($i = 2, j = 1$); and (4) high elevation, without exposure ($i = 2, j = 0$). The domain and four strata were then mapped onto the administrative level of ward in Tanzania and LC3 in Uganda. Wards and LC3s were designated as high elevation if a higher proportion of the area in the unit was above 1,200 m.a.s.l. Wards and LC3s were the primary sampling units (PSUs).

Given budget constraints, the total number of PSUs was fixed at 40, with half distributed in exposed and the other half in nonexposed areas, so that village-level characteristics might also be compared. The 40 PSUs were then allocated between the two elevations and two countries proportionate to the probability of selection, and drawn using systematic random sampling from a list frame with a random start.

The final sample in Uganda consists of 27 PSUs, of which 18 are located in nonexposed and 9 in exposed areas. In Tanzania, out of the 13 selected PSUs, 11 are located in exposed and 2 in nonexposed areas. The spatial representation of the primary sampling units is shown in Figure 2.1 with sampled sites highlighted. Figure 2.2 shows only the sites surveyed.

The secondary sampling unit (SSU) in both Uganda and Tanzania was a village, with a probability of selection that varies by PSU. Whether a community selected in the sample had been properly classified as exposed or nonexposed was then verified at the site. Within each village, a constant sample size of 20 households was maintained across villages, largely for convenience in monitoring the work of enumerators, who were posted in the villages for 1 year. Households were selected from a current village listing. The overall probability of selection is unique to each household, and

Figure 2.2 Sites sampled for survey

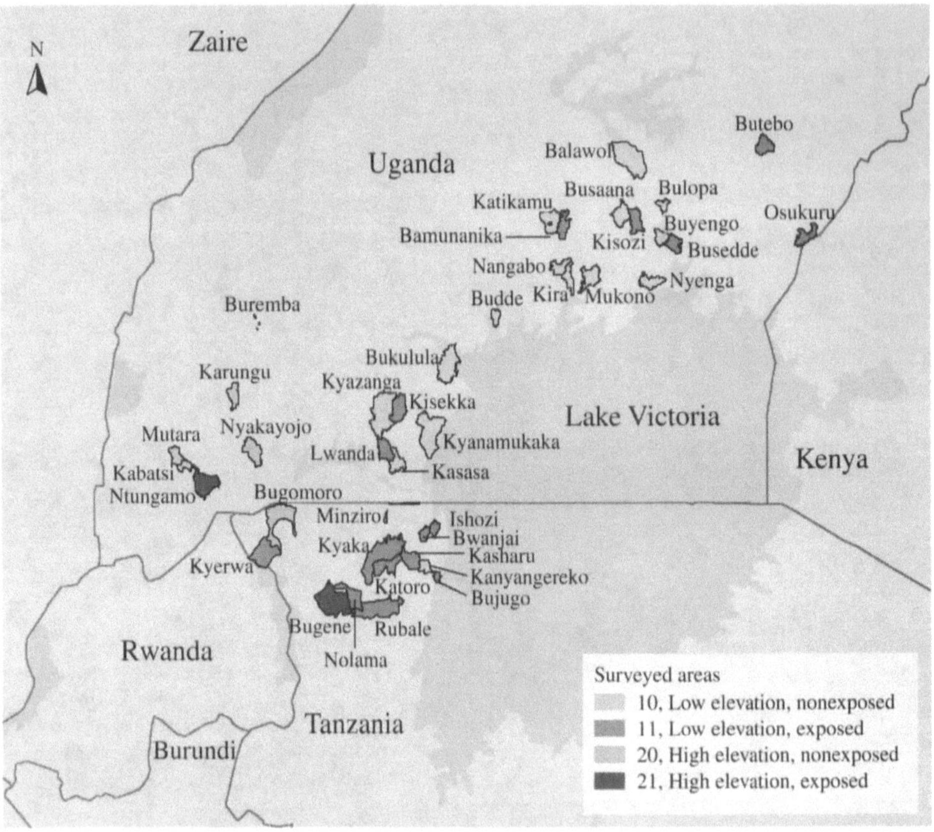

is defined as the product of the sampling fractions at each level. Weights have been used in the descriptive analyses presented in this report.

Industry Analysis

The first step in the sectoral analysis was to identify the range of technology options, through expert and stakeholder consultations,[1] based on (1) the size of current and expected economic losses associated with specific production constraints and (2) the likelihood that new technologies to mitigate these losses, and that are also suited to the Ugandan context, can be generated.

Five key production constraints were identified: wilts,[2] weevils, nematodes, black Sigatoka, and soil fertility decline. All five lead to productivity losses on affected banana mats, if appropriate interventions are not made. In some cases, declining productivity leads to significant reduction in the productive life cycle of affected mats, and consequently, of the grove or plantation as a whole.

To address this feature, the analytical model was formulated as in Figure 2.3. The

[1] This included consultations with research managers and banana experts from Uganda and international research centers.

[2] Both *Fusarium* and bacteria wilt were originally identified, but *Fusarium* wilt was later omitted from the analysis to allow more attention to data collection and analysis for the rapidly growing interest in assessing the potential economic threat of banana bacteria wilt.

Figure 2.3 Graphical depiction of analytical model used to assess the economic benefits of mitigating productivity losses

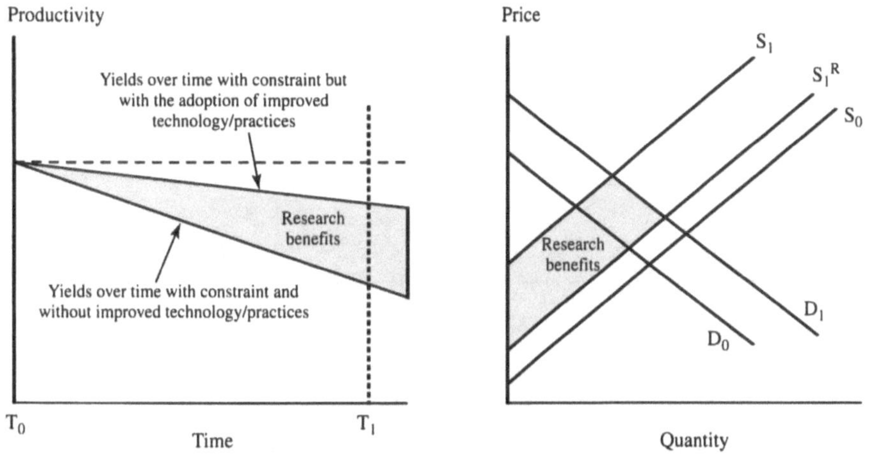

Notes: S_0 and D_0 are the initial supply and demand situations at time T_0, S_1 is the supply situation at time T_1 without research, S_1^R is the supply situation at time T_1 with research, and D_1 is the demand at time T_1.

left panel illustrates the general case of productivity decline over time at a specific mat or plot. Without intervention, or with ineffective intervention, yields tend to decline over time—perhaps dramatically so. The use of recommended practices or technologies might mitigate, but usually cannot eliminate losses. The "with technology" case is shown in the upper curve that declines more slowly, generating an economic payoff for the farmer relative to the "no technology" case. The economic payoff is represented by the shaded area between the curves, which increases over time. At some point productivity may become so low that banana production is no longer economic at such sites.

In the right panel of Figure 2.3, the cumulative effect of productivity losses on the industry, in the absence of effective interventions, is represented by the shift in the supply curve from S_0 to S_1. The availability and use of improved technologies and practices across the industry can lead to a supply situation depicted by S_1^R. In the absence of other supply changes (for example, an expansion in banana cultivation area), S_1^R would lead to an outcome that is inferior (less supply at higher prices than S_0), but substantially better than the counterfactual, S_1. The size of research benefits to the industry is shown by the shaded area.

Evaluating the economic effects of technical change involves accounting for the timing of the cost-and-benefit stream associated with the generation, diffusion, and use of each new technology. Particularly in the case of biotechnology (for both scientific and regulatory reasons), there may be lengthy lag times between the initial research investment costs and the eventual adoption of research-generated technologies and, hence, to a stream of research benefits. Figure 2.4 represents schematically the timing of flows of benefits and costs from a successful investment in developing a new technology.[3] The vertical axis represents the flow of benefits and costs in a particular

[3] Many new technologies are not successful in the sense that they are never developed for commercial use or adopted in the field. The figure refers to a new technology that is successful and adopted.

Figure 2.4 Time profiles of research costs and benefits

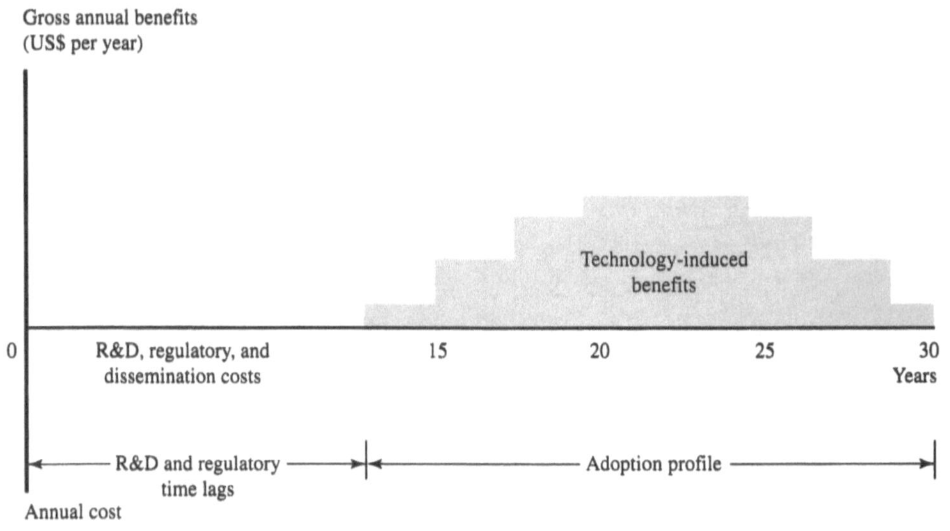

Source: Alston, Norton, and Pardey (1995).

year and the horizontal axis represents years after the commencement of the R&D investment.

Initially, R&D involves expenditure without benefits so that, during the R&D phase and possibly the regulatory phase, only costs (negative benefits) are incurred. Even when a viable product is available for commercial distribution, there are further time delays before the peak level of adoption of the new technology is achieved. The mix of adoption conditions across an entire industry drives the shape of the overall benefit profile, but the benefits derived from a technology gradually diminish as the technology becomes obsolete (that is, as better technologies are developed and adopted), depreciates (for example, as pests evolve to bypass host-plant resistance), or becomes uneconomic for other reasons.

In Chapter 10, IFPRI's DREAM model is applied using this framework.[4] Parameters include supply conditions, the potential supply-shifting effect of new technologies, and expected growth in banana demand (D_0 to D_1 in Figure 2.3), driven primarily by population growth. Application of the DREAM model requires an explicit specification of the market, policy, production, and adoption conditions likely to influence the generation, use, and hence, impact. We define these bundles of conditions as technology scenarios. Simulations are then used to generate time streams of potential benefits for each scenario.

Conclusions

The overall design of this research represents an attempt to address the multiple dimensions of technical change. At the farm level, the adoption of banana hybrids, transgenic cultivars, and best management practices was investigated, as well as the technical efficiency of banana production, taking soil and labor constraints into account. Best management practices include those designed to enhance soil fertility, and these practices tend to be labor intensive. The underlying conceptual framework for

[4] See Alston, Norton, and Pardey (1995, 386–394) for a complete description and specification of the model.

these analyses is the theoretical model of the household farm, which is particularly suitable for investigating farmer decision-making in situations where the costs of transacting in markets can be high. The economic surplus approach is applied at the industry level, and is aimed at providing key information to support policy decisions concerning investment priorities. The corresponding analyses and more details regarding each model are presented in separate chapters of Part III.

The sample design represents an effort to address the challenge of defining the "counterfactual." The research team developed a sampling frame that controls for previous exposure to new banana cultivars and practices. The sample of villages and households is also stratified by elevation, a parameter that is highly correlated with pests and disease pressures, as well as soil fertility, which are major constraints to banana productivity and the focus of banana improvement efforts.

References

Alston, J. M., G. W. Norton, and P. G. Pardey. 1995. *Science under scarcity: Principles and practices for agricultural research evaluation and priority setting.* Wallingford, U.K.: CAB International.

de Janvry, A., M. Fafchamps, and E. Sadoulet. 1991. Peasant household behaviour with missing markets: Some paradoxes explained. *Economic Journal* 101 (November): 1400–1417.

Duflo, E., and M. Kremer. 2003. Use of randomization in the evaluation of development effectiveness. Paper prepared for the World Bank Operations Evaluation Department, Conference on Evaluation and Development Effectiveness, 15–16 July, Washington, D.C.

Feder, G., and D. Umali. 1993. The adoption of agricultural innovations: A review. *Technological Forecasting and Social Change* 43: 215–239.

Feder, G., R. Just, and D. Zilberman. 1985. Adoption of agricultural innovations in developing countries: A survey. *Economic Development and Cultural Change* 33 (2): 255–298.

Heckman, J. J. 1976. The common structure of statistical models of truncation, sample selection, and limited dependent variables and a simple estimator for such models. *Annals of Economic and Social Measurement* 5 (4): 475–492.

———. 1990. Varieties of selection bias. *American Economic Review* 80 (2): 313–318.

Johnston, J., and J. di Nardo. 1997. *Econometric methods.* Fourth edition. New York: McGraw-Hill.

Kerr, J., and S. Kolavalli. 1999. *Impact of agricultural research on poverty alleviation: Conceptual framework with illustrations from the literature.* Environment and Production Technology Division Discussion Paper 56. Washington, D.C.: International Food Policy Research Institute.

Meinzen-Dick, R., M. Adato, P. Hazell, and L. Haddad. 2007. Findings of an integrated economic and social analysis. In *Agricultural research, livelihoods, and poverty: Studies of economic and social impacts in six countries,* ed. M. Adato and R. Meinzen-Dick. Baltimore: Johns Hopkins University Press.

Speijer, P. R., C. S. Gold, E. B. Karamura, and J. N. Kashaija. 1994. Banana weevil and nematode distribution patterns in highland banana systems in Uganda: Preliminary results from a diagnostic survey. In *African Crop Science Conference proceedings,* ed. E. Adipala, M. A. Bekunda, J. S. Tenywa, M. W. Ogenga-Latigo, and J. O. Mugah. Kampala: African Crop Science Society.

Tushemereirwe, W., D. A. Karamura, H. Sali, D. Bwamiki, I. Kashaija, C. Nankinga, F. Bagamba, A. Kangire, and R. Sebuliba. 2001. Bananas (*Musa* spp.). In *Agriculture in Uganda,* Vol. 11, *Crops,* ed. J. Mukiibi. Kampala: Fountain Publishers/Technical Centre for Agricultural and Rural Cooperation/National Agricultural Research Organization.

Part II. Research Context

CHAPTER 3

Overview of the Banana Economy in the Lake Victoria Regions of Uganda and Tanzania

Robert Kalyebara, Jackson M. Nkuba, Mgenzi Said Ramadhan Byabachwezi, Enoch Mutebi Kikulwe, and Svetlana Edmeades

This chapter summarizes the major features of the banana economy in the Lake Victoria region of Uganda and Tanzania, describing the economic importance of the crop, banana types and uses, and production and economic constraints. The East African highlands are recognized as a second center of banana diversity (the first being the Indo-Malaysian region of Asia), and most cultivars grown are endemic to the region. The numerous cultivars are distinguished by genome, phenotype, and use. Biotic pressures and declining soil fertility have caused major production problems in recent decades. A major food staple, bananas are also traded in local markets, but their perishability and transport costs contribute to risks for traders and to difficulties linking supply areas with areas of strong demand. Processing opportunities have not been fully exploited. Exports are largely regional or directed toward a niche market of emigrants. Tastes and preferences in the region favor the endemic types compared with sweet dessert bananas most often demanded by consumers on the world market.

Economic Importance

The banana is a major food staple in both Uganda and Tanzania, as well as an important cash crop in the local economy. An indispensable part of life in Uganda, bananas provide an estimated 30 percent of the calories, 10 percent of the protein, and 5 percent of the fat intake of the population, representing 25 percent of the total value of agricultural output (FAO 2002). Per capita annual consumption of bananas in Uganda is the highest in the world at approximately 0.70 kg per person per day (INIBAP 2000; NARO 2001). Compared with other important crops in the country, banana occupies the greatest proportion of utilized agricultural land (about 1.4 million hectares, or 38 percent of total utilized land), making it the most widely grown crop. It serves as one of the most important food security crops for Central, Western, and Eastern Regions of Uganda (NARO 2001).

Similarly, in Tanzania, bananas are the staple food of an estimated 20–30 percent of the total population (Walker, Hebblethwaite, and Bridge 1984). In the heavily banana-based farming systems, such as Kagera and Kilimanjaro regions, about 70–95 percent of households grow bananas for food and/or cash, and the average field size of the bananas grown ranges from 0.5 to 2.0 ha per household (Byabachwezi and Mbwana 1999). In other areas, only a few banana plants are maintained by households, mainly for dessert and roasting. Banana ranks first as a major food staple, and second or third as a cash crop in the banana- and/or coffee-based farming systems of Kagera, Arusha, Kilimanjaro, and Mbeya regions (Nkuba et al. 2003).

Banana Types and Uses

Most of the banana cultivars grown in the Lake Victoria region of Uganda and Tanzania are endemic to (or consistently present in) the East African highlands. The East African highlands is recognized as a center of banana diversity. NARO (2001) estimates that as much as 85 percent of the bananas grown in Uganda are EAHBs. The endemic banana cultivars (AAA-EA genomic group) consist of two use-determined types: cooking bananas (used to make *matooke*) and beer bananas (*mbidde*). In Uganda, these bananas have been classified by Karamura and Pickersgill (1999) into five clone sets according to phenotype characteristics (observable characteristics determined in part by genotype, in part by environment, and in part by the interaction of genotype with environment). The five clone sets are named Musakala, Nakabululu, Nakitembe, Nfuuka, and Mbidde. Farmers across the Lake Victoria region of Uganda and Tanzania distinguish cultivars based on a subset of these phenotypic characteristics and use.

Nonendemic bananas grown in this region include farmers' cultivars introduced from other regions of the world (here, we refer to these as "exotics") and a number of recently bred hybrids developed by breeders at FHIA in Honduras and the IITA in Nigeria. Exotic cultivars include beer and sweet bananas (AB, ABB, and AAA genomic groups) and roasting bananas or plantains (AAB genomic group). Hybrids are tetraploids (AAAA, AAAB, and other tetraploid genomic groups). Figure 3.1 summarizes this information by type of cultivar, genomic group, and most common use.

The total number of banana cultivars in East Africa has not been determined, and it is a subject of on-going debate among breeders. Each banana type or clone set has multiple cultivars that are known by local names. One source of confusion is that the same local name may be given to more than one clone, or a single clone may have several different names in different parts of the country. Karamura (1998) recognized 238 named banana cultivars growing in Uganda. Through her study of morphological variation and numerical analysis of these cultivars, synonyms were removed, leaving 84 distinct cultivars classified into five clone sets. Collections were also made throughout Kagera Region and part of southwest Uganda in the 1980s. High rates of synonyms and homonyms were evident in this first collection of 300 entries (Byabachwezi and Mbwana 1996), and in 1996, it was screened, reducing the number of distinct entries to 153 cultivars, mostly of the East African highland type (Mbwana, Byabachwezi, and Mkulila 1996).

The banana fruit and plant are used in many ways in the Lake Victoria region of Uganda and Tanzania, although there are some differences in consumer preferences in rural and urban areas. Consumption preferences for the staple food appear to be less specific to use type in Tanzania than in Uganda. Compared to the steamed, mashed banana (*matooke*) that is the staple dish in Uganda, the staple dish in Tanzania is a mixture of bananas, beans, and vegetables. The cooked food and juice also have cultural functions in some stages of wedding and funeral rites. The different parts of the crop have different uses in the daily life of a farm household. The leaves are used in the steaming of food, sheaths are used to make ropes and crafts, and pseudo-stems provide fodder. In Tanzania, cooking and brewing types are most prevalent in the Lake Zone, Northern Zone, and Southern Highlands. Dessert and roasting types are more popular in the Eastern Zone and Zanzibar. A single cultivar may have multiple and different uses within and between communities as a consequence of habits, taste preferences, and relative scarcity (Nkuba et al. 2003).

Figure 3.1 General classification of banana cultivars, genomic groups, and uses

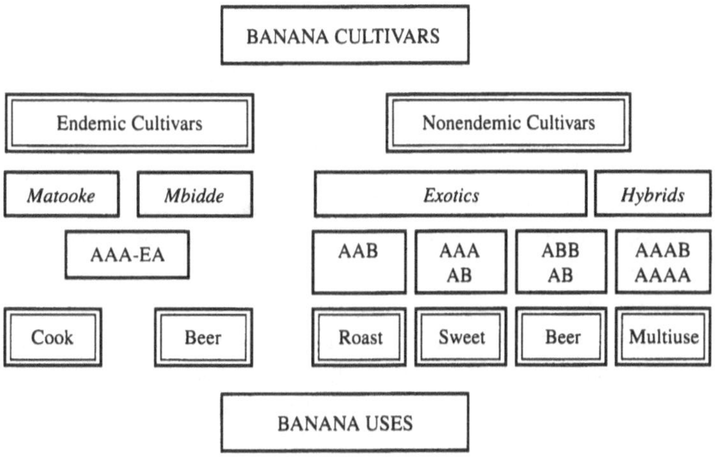

Source: Edmeades (2003).
Note: The genome group for Sukali Ndiizi, a small sweet banana, has been revised from AB to AAB.

Areas of Production

An estimated 61 percent of the Ugandan banana crop is produced in the Western Region, 30 percent is produced in the Central Region, and the remainder in the Eastern Region (UBOS 2003). Most banana production takes place on small subsistence farms of less than 0.5 ha using farming methods with low levels of external inputs (Gold et al. 1998).

In Tanzania, major zones of production are located in the highlands of Kagera and Mara regions in the Lake Zone, followed by Kilimanjaro, Arusha, and Tanga in the Northern Zone; Mbeya, Ruvuma, and Iringa in the Southern Highlands Zone; Coast and Morogoro in the Eastern Zone; and Zanzibar Islands. The Lake Zone has the highest banana production, followed by the Northern Zone. In 1997/98 and 1998/99, the Lake Zone produced about 45 percent and 43 percent of the total banana production in the country, respectively (MAC 2000).

Production Constraints

The life span of banana groves ranges from as low as 4 years in central Uganda to several decades. Speijer, Kajumba, and Tushemereirwe (1999) reported plantations as old as 30 years in western Uganda, and higher ages have been reported anecdotally. Until the past few decades, banana was considered a highly sustainable crop in Uganda, with long plantation life and stable yields. During the past 30 years, banana production patterns have been changing. Acreage has increased or remained stable in most of the Western Region, while declining in the Central and Eastern regions (Gold et al. 1998). This shift has been attributed to the increasing severity of production constraints. In particular, declining soil fertility, pests, and diseases severely reduced production in some areas.

Included among the most widespread biotic problems are weevils; black Sigatoka disease; Panama disease or *Fusarium* wilt; and banana bacteria wilt, which causes yield losses as high as 80 percent. Nematodes are a major problem in some areas. Different levels of susceptibility among cultivars have been observed. Weevils are insects that attack all types of banana cultivars, although the intensity of weevil damage has been found to decrease with elevation (Gold et al. 1994). Black Sigatoka is an airborne fungal disease, which affects endemic and some exotic cultivars. *Fusarium* wilt is another fungal disease that attacks the roots of banana plants. The exotic brewing cultivars are particularly susceptible to it (Gold et al. 1993). Bacteria wilt has emerged as a new and major disease since 2001. Due to its severity, and the fact that presently there is no cultivar in Uganda that is resistant to the disease, it poses a major threat to banana production in the country. Therefore, research priorities have recently shifted to mainly finding a solution to control banana bacteria wilt.

Drastically declining yields in the historical areas of production have led to the replacement of bananas with annual crops, and the locus of banana production has shifted to the southwest, where productivity biotic pressures are less, but the distances to urban markets are greater. Productivity in central Uganda is estimated at 6.0 ton/ha, while in the southwest it is 17 ton/ha, still low compared to the potential 60 ton/ha attainable at research stations (Tushemereirwe et al. 2001). Despite the decline in production, banana is still the most preferred staple in many localities, commanding a relatively high price in urban markets.

In Uganda, available historical data reveal sharp declines in both output and yields between 1970 and the early 1980s, followed by stagnating national yields (FAO 2004). The decrease in production in the 1970s and 1980s resulted from a severe outbreak in banana weevil in the southwest (for example, Masaka) and a combination of increasing weevil pressure and declining management and soil fertility in the Central Region, then the locus of banana production. Increasing banana production between 1980 and 2003 is largely due to area expansion and the shift in production to more productive regions in the west. Evidently, banana yields have not recovered to pre-1980 levels, despite intensifying efforts to improve productivity through R&D.

Soil fertility depletion is a major contributor to declining banana yields in Uganda (Gold et al. 1998). Either inorganic or organic fertilizers can be used to improve soil fertility on small farms. There is very little use of chemical fertilizers in banana plots. The low profitability of inorganic fertilizers explains their low adoption by farmers, and major improvement in market infrastructure is a prerequisite for substantial adoption to occur (Nkonya et al. 2005). Organic fertilizers are the most important fertility amendments that farmers apply to cropland. Intensive application of organic fertilizers in the form of mulch (that is, grass, crop residues, or kitchen refuse) or animal manure (that is, cattle, goat, pig, and poultry manure) in banana groves can improve and maintain soil fertility even when fertility is inherently low.

The efforts of Ugandan banana researchers during the 1990s have been directed toward formulation of operational strategies to address pest and/or disease problems and create more awareness of the importance of fertility management by farmers. Although a number of crop-management techniques have been developed and recommended to farmers over the years, most are practiced irregularly and sometimes not at all. Not practicing these techniques encourages bare soils between mats, where erosion starts too readily. Farmers have had decreasing access to grass mulch and manure in many (but not all) areas. The direct consequences are low yields and reduced longevity of the banana plantation.

In addition to soil and biotic constraints, banana productivity and returns to production inputs are further limited by the following socioeconomic constraints: (1) high cost of production due to high intensity of demand for labor, manure, and mulch; (2) temporally limited markets, especially at peak harvest, as a consequence of the high perishability of the product; (3) high costs of marketing fresh bananas due to their bulk, perishability, and long haulage distances to markets; and (4) the inability of most farmers to access markets and exploit opportunities.

Particularly in the Central Region, the ability of farmers to respond to the major biotic and abiotic constraints is limited due to widespread poverty, very limited use of commercial inputs, and lack of cultivars that are resistant to pests and diseases. Efforts to contain the spread and impact of these biotic pressures on productivity have concentrated on scaling up dissemination of currently available best management practices and of introduced hybrid cultivars with tolerance to banana weevils, black Sigatoka, and *Fusarium* wilt. Introductions of hybrids into Uganda during the 1990s targeted areas where pest and disease pressure is relatively high and yields of farmers' cultivars were declining rapidly. In Uganda, preferred endemic cultivars were also selected for dissemination through a major program to promote the use of "clean" planting materials (planting materials that are free of pests and diseases). Current research efforts are primarily directed at breeding cultivars with resistance to banana bacteria wilt, black Sigatoka, weevils, and nematodes; and developing cultural and biological control techniques for major pests and diseases. Banana improvement by means of conventional plant breeding techniques (using natural pollination by insects or bats, or by deliberately moving pollen from one of the male flowers of one clone to female flowers of another) has proved extremely difficult (see Chapter 4).

In the major banana-producing areas of Tanzania, available data also indicate that production has remained largely stagnant or declined during the 1990s (MAC 2000; Nkuba et al. 2003). Major banana-production constraints faced by farmers in Tanzania are similar to those faced by farmers in Uganda and include declining soil fertility, increased pressure of pests, and poor agronomic practices. The major pests of bananas found in Tanzania are banana weevils (*Cosmopolites sordidus*), nematodes (various species), black Sigatoka (*Mycosphaerella fijiensis*) and *Fusarium* wilt (*Fusarium oxysporum* cv *cubence*).

Many banana producers in the Lake Victoria region of Tanzania are unaware of good banana-crop husbandry, because those who migrated to banana-growing areas often are not banana growers by culture and are unfamiliar with traditional practices. Others, who are well aware of practices, are discouraged by low returns to labor, given current market prices and the yield levels of endemic cultivars.

Local Markets

About 90 percent of Uganda's banana output is consumed in the domestic market; the remainder is exported as either fresh or processed products. The importance of local demand is illustrated by the fact that Uganda has the highest per capita annual consumption of bananas in the world. Most of the output is consumed as fresh bananas. In the main banana-producing areas in the west, recent survey data show that approximately 65 percent of output is consumed on-farm, while only 35 percent is sold (Chapter 8). In less productive areas, households supplement home-produced banana output by consuming other starchy staples, such as maize, cassava, and sweet potatoes.

Growers who sell bananas typically sell them at the farm gate to traders for resale in urban markets. Future prospects for local banana markets appear to be good, given the growing population and upward price trend in Kampala and other cities. In Uganda, the main constraint limiting the profitability of banana marketing is the high cost of transportation from major suppliers to Kampala and the risks involved. Transport costs account for 80 percent of total marketing costs (NARO 2005). Because of high costs and wide fluctuations in banana prices, marketing margins vary widely during the year (Table 3.1). The distribution of marketing margins for cooking bananas shows that brokers and agents earn the highest margins per bunch of banana sold (about 50 percent of the farm-gate price). Purportedly, brokers are better organized and strike

Table 3.1 Net marketing margins per bunch of cooking bananas from western Uganda to Kampala (Ush)

Agent	Average	Minimum	Maximum
Broker/agent	812	200	1,800
Rural retailer	479	100	1,100
Wholesaler/transporter	633	200	2,933

Sources: Banana marketing baseline survey; NARO (2005).

good deals because they exploit information asymmetry in the market and poor organization of the farmers.

In Tanzania, major local markets of bananas are found in urban areas located outside banana-producing zones, including Mwanza and Shinyanga for bananas produced in the Lake Zone, and Dodoma and Dar es Salaam for bananas produced in Northern, Eastern, and Southern Highland zones (Figure 3.2).

Nkuba et al. (2003) conducted a study of banana markets covering major banana-growing zones and urban areas of demand (including Bukoba, Mwanza, Shinyanga, Musoma, Dodoma, Arusha, Moshi, Singida, Morogoro, Dar es Salaam, Zanzibar, Iringa, and Mbeya). In each zone of production, farm-gate prices for banana bunches fluctuate within and between harvesting seasons. Large traders receive the largest shares of gross marketing margins, followed by retailers, small traders, and farmers. The share of gross marketing margins received by the last two groups varied from as little as 6.6 percent in Kagera Region to as much as 30 percent in Morogoro. This variation reflects the fact that farmers and small traders have no access to market information in the areas of high consumer demand, and hence, little bargaining power.

The costs of transporting bananas from the areas of supply to the nearest wholesale market depend on access to roads and vary substantially among the production zones. At the time of the 2002 survey (Nkuba et al. 2003), transport costs were relatively high in the Lake Zone (TSh 2.50–3.75 per bunch/km) compared to the Northern (TSh 1.50) and Southern Highlands (TSh 1.65) zones. In major Tanzanian markets, an upward trend is visible from 1991 to 2000 in the retail prices per bunch of cooking and dessert bananas, with large fluctuations during the past 5 years (Figure 3.3).

In the near future, the growing population could have a positive effect on the national demand for bananas, depending on the prices of substitute starchy staples and cross-price elasticities. As incomes rise, demand will also depend on income elasticities. One critical factor that will contribute to effective demand is improvement in main and feeder roads, so that supply and demand areas can be reached easily by traders. Existing road infrastructure does not connect traders easily to major supply areas. Many producers have no information about prevailing prices in markets. Construction of tarmac and feeder roads will alleviate this situation, though there are other problems: Banana bunches are bulky in nature and perishable. Farmers have little alternative but to sell bananas at "throw-away" prices. Banana traders have no insurance coverage and are obliged to charge high marketing margins to offset risks.

A second major factor affecting national demand for bananas, and one which may eliminate some of these problems, is application of processing technologies. Processing of bananas into products with longer shelf life, such as dried bananas, banana flour, banana biscuit, and breads, is not well developed in Tanzania. Only an estimated 5 percent of all banana production is pro-

Figure 3.2 Map of Tanzania showing banana supply routes

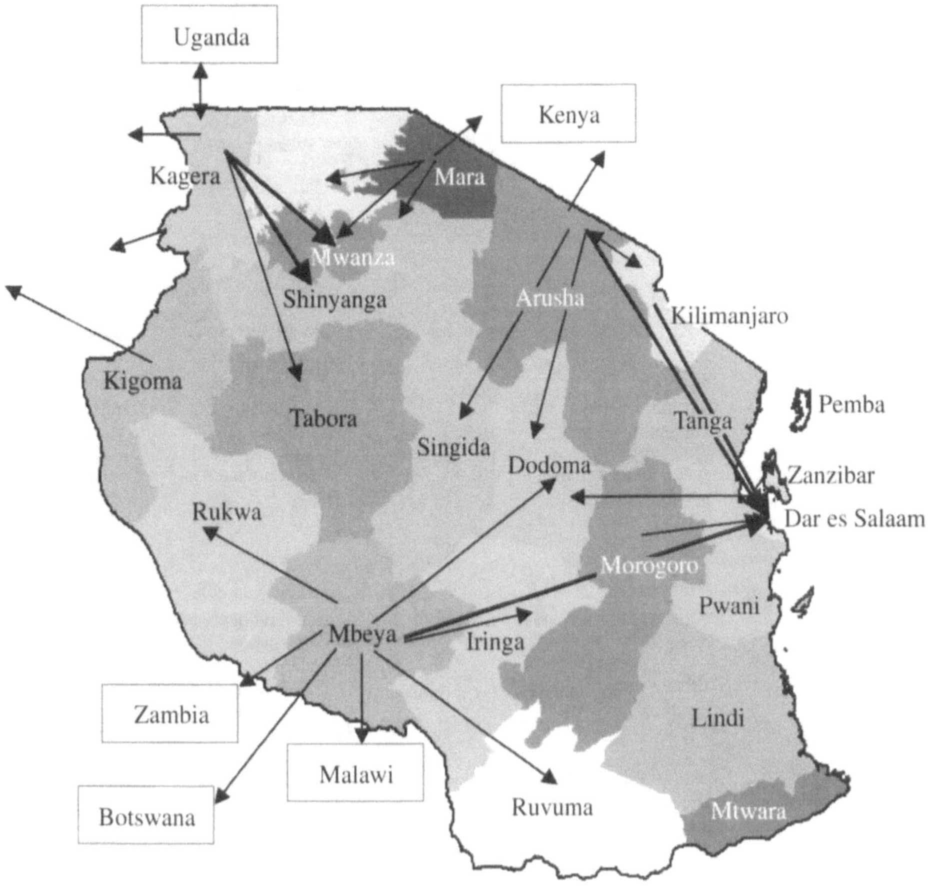

Source: MAFS (2001).

cessed. Principal buyers for cooking bananas are consumers. Buyers for roasting and frying bananas are mainly business enterprises (such as hotels and bars) that sell the processed/roasted bananas (Nkuba et al. 2003). Some banana types and their products are exclusively demanded within local banana-growing areas, whereas others are demanded within and/or outside these areas (Table 3.2).

Exports

Uganda is the second-largest producer of bananas in the world, but it is among the smallest exporters. Most of its production is consumed nationally, with some regional trade and very small quantities exported to Europe.

The quantity of bananas crossing the border from Uganda into Kenya, Rwanda, and Congo, estimated from a 1999 survey at selected border points, is shown in Table 3.3. The total weight of bananas exported was estimated at 3,821 tons, with an approximate value of USh 600 million (US$400,000). Exports can be expected to increase in future years if demand for food increases in these countries due to population growth. A supply deficit of bananas and food shortages in neighboring counties assure Uganda of a market for bananas and other food staples.

Table 3.2 Banana product segmentation in Tanzania

Product	Demand locus	Rate of growth in demand	Chief competitors	Principal buyers	Market requirement
Cooking type	In all production and urban areas	High	Maize, rice, cassava, and sweet potatoes	Traders and consumers with medium to high incomes	Efficient transportation system and good handling facilities for bananas
Brewing type	Within production areas	Medium	Industrial brews, beers, and soft drinks	Local people	Production of good quality cultivars
Dessert type	In both rural and urban areas of Mwanza, Shinyanga, Dar es Salaam, Dodoma, and Singida	High	Other fruits, such as mangoes, oranges, and mandarins	Traders and consumers (hotels, restaurants, and the like)	Improvement of transportation facilities
Roasting type	In urban areas in bars and pubs	Low	Round potatoes	Traders and beer drinkers	Increases in amount supplied and improvement of roasting techniques
Local beer	In rural areas of banana production	Medium	Industrial beers and soft drinks	Low-income people	Improvement of quality
Banana wine	In urban areas of major towns	Medium	Industrial beers and soft drinks	Medium- and high-income people	Standardization of quality and mass production
Banana biscuits, flours, breads, doughnuts, and the like	In areas not producing bananas and distant markets	Low to medium	Cereal products from maize, wheat, and rice	Traders and consumers	Improvement of processing technologies; product differentiation

Source: Nkuba et al. (2003).

Minor quantities of cooking bananas and small sweet bananas are also exported to Europe. Figure 3.4 shows that export quantities have gradually increased from 500 tons in 1994 to about 1,600 tons in 2002.

The combined value free on board (FOB) of exports to Europe of endemic cooking bananas and small sweet bananas (apple banana, or Sukali Ndiizi) in 1999 was approximately US$796,000 (ADC/IDEA Project 2001). The potential for growth in cooking banana exports to Europe is low, because this type is largely consumed by Ugandans living in Europe. Buyers do not expect it to become popular with European consumers, and investment in the export of cooking bananas is not recommended.

There is potential for expanding the market for small sweet bananas, but to accomplish it, major improvements would be needed in production systems to produce fruit that meets E.U. standards for quality and shelf life. Sukali Ndiizi is an exotic type currently produced almost entirely by smallholders for the domestic, regional, and European supermarkets. Research is underway to improve packaging and handling systems and to ensure high quality and long shelf life. Expansion will also depend on investment in packing facilities and refrigeration that meets E.U. food safety requirements. Such requirements mean that smallholders

Table 3.3 Estimates of cross-border trade between Uganda and neighboring countries, 1999 (tons)

Type of Banana	Lwakhakha	Malaba	Busia	Katuna	Mutukula	Total
Cooking	480	240	96	2,400	–480	2,736
Small sweet	96	48	48	5	0	197
Large sweet	240	576	72	0	0	888
Total	816	864	216	2,405	–480	3,821

Source: ADC/IDEA Project (2001).

Figure 3.3 Nominal retail prices (Tsh) of cooking and dessert bananas, 1991–2000

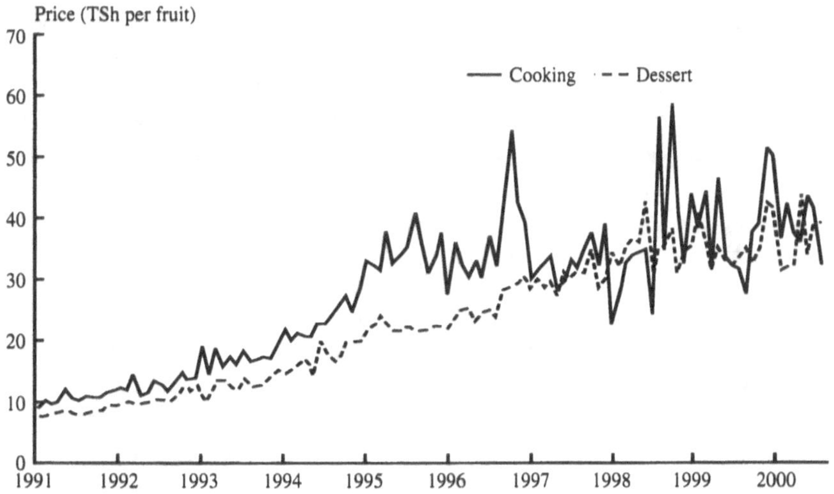

Source: MAFS (2001).

Figure 3.4 Uganda banana exports

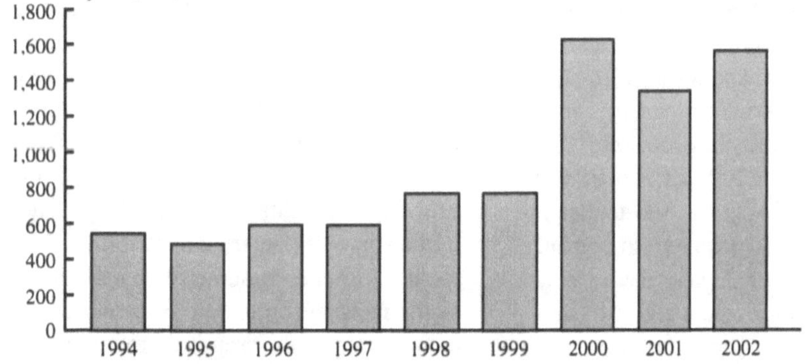

Sources: ADC/IDEA Project (2001); UBOS (2003).

would only be able to benefit from banana exports as contract growers for well-financed plantations with modern packing facilities, implying a fundamental change in the mode of production.

Most importantly, viable market opportunities are restricted to supplying niche markets for organic apple bananas and ethnic consumers of cooking bananas in Europe, the United States, and Asia. This situation reflects the fact that the country does not produce the Cavendish banana that is internationally traded, but a unique type of banana that is only consumed within the region (or by emigrants). Other developing countries, as well as Uganda, still face insurmountable hurdles to penetrate the biggest banana market: the market for Cavendish bananas. Uganda's low comparative advantage stems from prohibitive freight costs due to being landlocked, and poor quality compared to more efficient and larger suppliers in Latin America and the Caribbean.

Unlike Uganda and Rwanda, which export bananas (albeit in minor quantities) to European countries by air, Tanzania exports bananas only by road to neighboring countries, including Kenya, Uganda, Zambia, Malawi, Rwanda, Burundi, and Botswana. To date, there are no banana traders (individual, associations, or companies) operating at a large international scale in the country. In 2002, the estimated annual exports of cooking and dessert bananas were 3,230 tons sold in Kenya; 780 tons sold in Malawi, Zambia, and Botswana; and 470 tons sold in Burundi and Rwanda. Importation of bananas is uncommon, occurring primarily in communities scattered along the national borders. For export markets to be developed in Tanzania, substantial investment would be required to enhance postharvesting technologies, facilitate access to market information and infrastructure, and form marketing associations. As in Uganda, commercial farmers would need to orient their choice of planting materials toward the provision of attributes demanded by consumers in other countries, which are likely to differ substantially from those preferred locally. Even then, the world market is dominated by the Cavendish banana, and market entry would not be easy.

Conclusion

The crucial role of bananas in the national and local economies of Uganda and Tanzania, as sources of both food and cash, cannot be overemphasized. The East African highlands are also an important center of diversity for a unique type of banana. Banana diversity is reflected in multiple local uses. To some extent, the different genomic structure of cultivars grown by farmers confers differential genetic resistance to a number of biotic stresses. Nevertheless, biotic pressures have had a pronounced impact on yield levels and the geographical distribution of production, especially during the past three decades. Relieving these constraints through a combination of breeding and crop management strategies has therefore become a national priority (see Chapter 4). Banana production is also labor intensive, so that the development of local labor markets has implications for productivity (see Chapter 8). In addition to the market access constraints faced by many smallholder farmers in Africa, the properties of bananas, such as their bulkiness and perishability, pose challenges for the marketing system. Processing technologies to increase shelf life could facilitate product differentiation and marketing, and improved road infrastructure will augment local trade, generating rural income. International trade is, so far, negligible. The uniqueness of the East African highlands cooking banana has implications for international trade. First, there is limited consumer demand outside of the region for this type of banana. Second, as long as the crop is a preferred food staple, there is need for regional self-sufficiency.

References

ADC/IDEA Project (Agribusiness Development Centre/Investment in Developing Export Agriculture Project). 2001. Apple banana. ADC Commercialisation Bulletin 8. Kampala: U.S. Agency for International Development. Available at www.foodnet.cgiar.org/inform/Idea/Banana.PDF.

Bank of Uganda, Research Department. 2000. *Quarterly economic report.* Kampala: Bank of Uganda.

Byabachwezi, M. S. R. 2002. Banana research in Kagera Region: Situation in its social context. Manuscript.

Byabachwezi, M. S. R., and A. S. S. Mbwana. 1996. *Characteristics of East Africa highland bananas and their cross-naming in Tanzania.* Dar es Salaam, Tanzania: Commission for Science and Technology.

———. 1999. Banana and its improvement efforts at Maruku. In *Planning the future: Past, present and future perspectives of land use in the Kagera Region,* ed. E. C. R. Folmer, C. Schouten, and F. P. Baijukya. Bukoba, Tanzania: Agricultural Research and Development Institute, Lake Shore–Maruku.

Edmeades, S. 2003. Variety choice and attribute trade-offs within the framework of agricultural household models: The case of bananas in Uganda. Ph.D. thesis, North Carolina State University, Raleigh.

Gold, C. S., M. Ogenga-Latigo, W. Tushemereirwe, I. N. Kasaija, and C. Nankinga. 1993. Farmer perceptions of banana pest constraints in Uganda: Results from a rapid rural appraisal. In *Biological and integrated control of highland banana and plantain pests and diseases. Proceedings of a research coordination meeting,* ed. C. S. Gold and B. Gernmill, Cotonou, Benin, November 12–14, 1991. Kampala: International Institute for Tropical Agriculture.

Gold, C. S., P. Speijer, E. Karamura, W. Tushemereirwe, and I. Kashaija. 1994. Survey methodologies for banana weevil and nematode damage assessment in Uganda. *African Crop Science Journal* 2 (3): 309–321.

Gold, C. S., A. Kiggundu, D. Karamura, and A. Abera. 1998. Diversity, distribution and selection criteria of *Musa* germplasm in Uganda. In *Bananas and food security,* ed. C. Picq, E. Fouré, and E. A. Frison. International symposium on the socioeconomics of non-export banana production, Douala, Cameroon, November 10–14, 1988. Montpellier, France: International Network for the Improvement of Banana and Plantains.

INIBAP (International Network for the Improvement of Banana and Plantains). 2000. *Annual report.* Montpellier, France: INIBAP.

Karamura, D. A. 1998. Numerical taxonomic studies of the East African highland bananas (Musa AAA–East Africa) in Uganda. Ph.D. thesis, University of Reading, Reading, England.

Karamura, D. A., and B. Pickersgill. 1999. A classification of the clones of East African highland bananas (*Musa*) found in Uganda. *Plant Genetic Resources Newsletter* 119: 1–6.

MAC (Ministry of Agriculture and Cooperatives). 2000. Basic data. Agriculture and livestock sector. Dar es Salaam, Tanzania: MAC.

MAFS (Ministry of Agriculture and Food Security). 2001. Unpublished data. Dar es Salaam, Tanzania: MAFS, Market Information Department.

Mbwana, A. S. S., M. S. R. Byabachwezi, and S. I. Mkulila. 1996. *Banana germplasm collection.* Bukoba, Tanzania: Agricultural Research Institute Maruku.

NARO (National Agricultural Research Organization). 2001. Multilocational banana germplasm evaluation trials: Third report. Dar es Salaam, Tanzania, NARO. Manuscript.

———. 2005. Banana marketing and utilisation in Uganda: Baseline survey report. National Banana Research Programme, NARO, Uganda. Manuscript.

Nkonya, E., J. Pender, C. Kaizzi, E. Kato, and S. Mugarura. 2005. *Policy options for increasing crop productivity and reducing soil nutrient depletion and poverty in Uganda.* Environment and Production Technology Division Discussion Paper 134. Washington, D.C.: International Food Policy Research Institute.

Nkuba, J. M., S. R. B. Mgenzi, M. Ishika, and C. Mushongi. 2003. *Evaluating the marketing opportunities for banana and its products in the principle banana growing countries of ASARECA. Case study of Tanzania.* Bukoba, Tanzania: Maruku Agricultural Research and Development Institute.

Speijer, P., C. Kajumba, and W. Tushemereirwe. 1999. Dissemination and adaptation of a banana clean planting material technology in Uganda. *InfoMusa* 8 (2): 11–13.

Tushemereirwe, W., D. A. Karamura, H. Sali, D. Bwamiki, I. Kashaija, C. Nankinga, F. Bagamba, A. Kangire, and R. Sebuliba. 2001. Bananas (*Musa* spp.). In *Agriculture in Uganda,* Vol. 11, *Crops,* ed. J. Mukiibi. Kampala: Fountain Publishers/Technical Centre for Agricultural and Rural Cooperation/National Agricultural Research Organization.

UBOS (Uganda Bureau of Statistics). 2003. *Statistical abstract.* Entebbe, Uganda: UBOS, Ministry of Finance Planning and Economic Development.

Walker, P. T., M. Hebblethwaite, and J. Bridge. 1984. *Project for banana pest control and improvement in Tanzania.* European Economic Community Report for Government of Tanzania. London: Tropical Development and Research Institute.

CHAPTER 4

Development and Dissemination of Improved Banana Cultivars and Management Practices in Uganda and Tanzania

Enoch Mutebi Kikulwe, Kephas Nowakunda, Mgenzi Said Ramadhan Byabachwezi, Jackson M. Nkuba, Josephine Namaganda, David Talengera, Enid Katungi, and Wilberforce Kateera Tushemereirwe

This chapter summarizes the major approaches employed to combat plant pests and diseases in the Lake Victoria region of the East African highlands. In Uganda, national researchers have pursued a short-term strategy of assembling endemic and nonendemic germplasm for evaluation, and a long-term strategy of breeding for resistance using a combination of crossing and genetic transformation. Practices targeting both management of the natural resource base and the crop itself are also recommended, because bananas demand intensive management to sustain yields. Planting material that is free of pests and diseases has long been a major concern for banana growers and national research institutions in both countries. Tanzania has no formal banana-breeding program, and although extension services in the Kagera Region have undertaken projects to eradicate pests and diseases with insecticides, they have been unsuccessful. As in Uganda, collections were made in search of resistant material, and new banana-management practices have been recommended. Through KCDP and the International Plant Genetic Resources Institute (IPGRI; now Bioversity International), a range of new materials—including the first banana hybrids—was introduced, tested, and disseminated. In Uganda, a number of dissemination mechanisms have been employed for both planting material and recommended practices. Given the reproductive properties of bananas, farmer-based, participatory methods have been recommended.

Research Priorities

The EAHB is susceptible to pests (nematodes, banana weevils) and diseases (black Sigatoka, banana streak virus) (Gold, 1998, 2000; Gold et al. 1998, 1999; Gold, Pena, and Karamura 2001). Traditionally, farmers replaced a diseased endemic cultivar by replanting suckers of the same cultivar obtained from their fellow farmers within or outside the community. Often, this process inadvertently contributed to the spread of the pests and diseases, because farmers did not recognize infested or diseased planting material or fully understand the life cycles and transfer mechanisms of pests and diseases, especially given continual evolution of new races and pathogens. In areas along the lake shores in Bukoba and Muleba districts of Tanzania, in particular, as a growing rural population intensified banana production to meet their food and cash needs, the performance of endemic banana cultivars deteriorated substantially. Mat numbers for some endemic cultivars diminished or they disappeared altogether from individual farms and/or communities.

The extension services of the Department of Agriculture in Tanzania were able to recognize banana production constraints and attempted to address them through emphasizing good crop husbandry, including the uprooting of diseased banana plants and use of insecticides. The impacts of their efforts were not impressive, however, given the high cost of insecticides and the difficulty for farmers to understand pest and disease complexes. Legacies from the colonial era did not improve farmers' trust in government recommendations. For example, farmers resisted the order of the colonial government to uproot all bananas infested by weevils, in part because of misunderstandings (Kabwoto 1974). Toward the end of the 1960s, an insecticide (dieldrin) was recommended for the control of banana weevils. The Bukoba Co-operative Union (BCU) supplied insecticides free of charge and during that time, 60 percent of the farmers applied the insecticide (Rald and Rald 1975). According to Bosch et al. (1996), after dieldrin application, farmers reported that banana plants fell over more frequently than before and complained that the insecticides "killed" their bananas. Most of the farmers have since been reluctant to apply chemical of any sort, including artificial fertilizers, in their banana groves (FSR 1989).

In Uganda, banana researchers identified host-plant resistance as one of the most feasible alternatives to control these problems (Ortíz and Vuylsteke 1994; Tushemereirwe et al. 2000), adopting a combination of short- and long-term strategies. The short-term strategy includes the assembly of endemic and nonendemic germplasm for evaluation and selection of resistant or tolerant cultivars, including the importation of hybrids from other breeding centers, such as FHIA in Honduras and IITA in Ibadan, Nigeria. The long-term strategy includes breeding for resistance in the National Banana Research Programme, NARO, Uganda. The objectives of the overall approach are to develop and disseminate banana genotypes that are resistant to local pests and diseases, have good agronomic characteristics, and are acceptable to end users. A time line identifying key dates and junctures in the history of banana research in Uganda is shown in Table 4.1.

In addition to work with germplasm, NARO has recommended improved banana-management practices. Bananas are produced by propagation continuously on the same piece of land, and therefore demand intensive management to sustain their yields. Both management of the natural resource base and those related with the crop itself are recommended. Natural resource management practices used in banana production include: (1) application of dry organic mulch (in the form of grass or crop residues), (2) manure (that is, animal waste and composting of household refuse), and (3) soil and water conservation bands. Farmers are also advised to carry out a number of other crop management practices to ensure good sanitation in their plantations so as to reduce pest and disease infestations and facilitate the management of their soil fertility. Sanitation practices include: (1) corm paring,[1] (2) detrashing (removal of dry leaves and sheath), (3) desuckering (removal of excess plants on

[1] In the traditional farming system, banana plantations are established with suckers of about 1.5 m high. The planting material is usually uprooted with its roots and soil and taken for planting in a new field. This method has disadvantages, because such planting materials can transmit pests (nematodes and weevils) and soil-borne diseases, such as *Fusarium* wilt, to the new field. Farmers have been trained to clean banana planting material by corm paring. Through paring, one is able to see tunnels formed by weevil larvae and necrosis due to nematodes. As an additional treatment, it was demonstrated to farmers that pared planting material can be dipped in hot water at a temperature between 52° C and 55° C for 20 minutes or in a pesticide solution to kill the deeply embedded pests (Gold et al. 1998; Tushemereirwe et al. 2003).

Table 4.1 Key dates in the history of banana improvement in Uganda

Activity	Institution	Year
Germplasm collection started	Kampala plantation	1920
	KARI	1940
Establishment of tissure culture (micropropagation)	Makerere University (banana)	1991
	KARI (coffee)	1993
Germplasm collection and evaluation re-established	Makerere University–Kabanyolo	1989
	KARI	1989
Banana micropropagation	KARI/IITA	1995
Germplasm characterization	KARI	1995–98
Germplasm evaluation for female and male fertility	KARI/IITA	1996–98
Cross pollination	KARI/IITA	1996
Banana embryo culture	KARI/IITA	1996
Cell culture	KARI	1998–2007
Taxonomic and screening for conventional breeding	KARI	2000
Field testing of plants from single cells (cell suspension)	KARI	2002–07
Biotechnology lab opened	KARI	2003

Sources: NARO/IITA annual reports; Makerere University, Faculty of Agriculture reports.
Notes: IITA is the International Institute of Tropical Agriculture; KARI is the Kawanda Agricultural Research Institute.

a mat), and (4) other residue management practices (such as stumping and corm removal). Splitting pseudostems and weevil trapping are also recommended. Details of the banana-management practices recommended to farmers are described in Tushemereirwe et al. (2003).

In Tanzania, although no formal plant breeding is undertaken, banana growers have also worked to counteract these pest and disease pressures by procuring new, clean planting material for endemic cultivars within and outside their communities, exchanging new cultivars with other farmers, and adopting banana cultivars introduced by public and private extension services. Furthermore, the extension service has sought to curb pest and disease pressures through training farmers on good banana-management practices. In mid-1997, the government of Tanzania and the Kingdom of Belgium cofinanced a major effort to propagate and diffuse superior banana cultivars under KCDP.

Selection and improvement strategies pursued by farmers, researchers, and extension agents are described next, following by a summary of dissemination mechanisms.

Figure 4.1 Introduction of exotic banana cultivars to Kagera Region by farmers

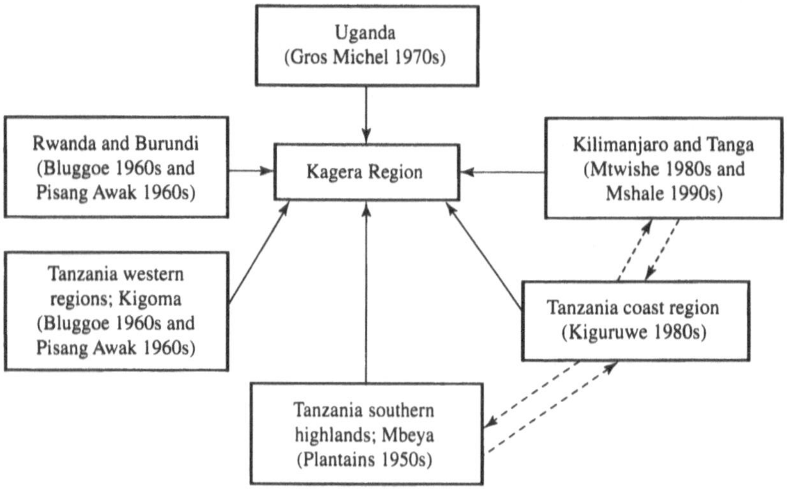

Source: Personal interviews with farmers in Kagera Region conducted by M. S. R. Byabachwezi and J. M. Nkuba.
Note: In each box, the name of the region is provided, along with the name of the cultivar and the approximate period of introduction in parentheses. The solid arrows indicate the original introduction of the exotic cultivars, and the broken arrows denote subsequent exchange of planting material across regions.

Germplasm Selection and Improvement Strategies

Farmer Introductions of Exotic Planting Material

Over the decades, farmers have continually sought pest- and disease-free planting material through exchange with farmers in other regions. For example, in Kagera Region, oral interviews with farmers suggest that a number of "exotic" (nonendemic, nonhybrids) cultivars grown today were introduced over the decades from Uganda, Rwanda, Burundi, and the coastal areas and southern highlands of Tanzania. These include Kijoge (Gros Michel, also known as "Bogoya" in Uganda), Kisubi/Kainja (Pisang Awak), Mtwishe (Medium Cavendish), Kiguruwe (Short Cavendish), Bluggoe, Plantains, and Mshale. Some EAHBs were taken from Kagera Region to other areas, where they are known as "Ndizi Bukoba" or "Ndizi Uganda." Grande Naine, another exotic cultivar, was introduced through unknown channels in 1993 (Figure 4.1).

The exotic cultivars appeared to withstand the most common banana constraints at that time and spread rapidly into areas where endemic bananas were especially affected by pests and diseases. These types then succumbed to other banana pests and diseases, such as *Fusarium* wilt (also known as "Panama disease"). Banana production again declined, especially in the eastern part of Bukoba and Muleba districts. In response, during the 1980s, farmers in Kagera Region diversified into other food crops, including maize, roots, and tubers.

Selection from Existing Germplasm

Apart from continuing efforts to develop new banana-management practices, aggravated production problems led Agricultural Research Institutes (ARIs) to begin the collection of banana germplasm at ARDI during the 1960s. Collections were made throughout Kagera Region and part of southwest Uganda, and by 1989, the collection numbered 153 unique entries (Mbwana, Byabachwezi, and Mkulila 1996).

Banana germplasm collections were made as early as the 1920s in Uganda. From

1990, diagnostic studies were conducted to identify and quantify banana production constraints. These studies were followed by collection of endemic germplasm with the aim of searching for cultivars resistant to the pests (weevils and nematode) and diseases (especially black Sigatoka). For the purposes of evaluating endemic materials for yield, pest, and disease reaction, two germplasm collections were established at Makerere University Agricultural Research Institute, Kabanyolo (MARIK), and Kawanda Agricultural Research Institute (KARI). Endemic germplasm was collected from farmers in different parts of the country.

Development of New Genotypes in Uganda

Formal banana breeding did not begin in Uganda until 1994. First, studies were conducted to identify fertile female banana cultivars (with potential to set seeds when pollinated) from the assembled set of endemic cultivars and pollen-fertile diploid clones to serve as sources of desirable attributes. A study of female banana fertility, conducted from 1996 and 1998, identified 12 of the most fertile EAHBs in the collection (Ssebuliba 2000). In addition to the low seed set in the endemic cultivars, a seed germination of only 1 percent was recorded when the seeds were sown directly into soil.

Embryo culture, which is now a routine practice in Uganda, is used to improve the germination of hybrid seeds derived from inter- and intraspecific crosses. The seeds are surface sterilized and aseptically cracked to expose the embryo. The embryo is extracted and inoculated onto an appropriate medium (Murashige and Skoog 1962) to achieve germination. Seedlings are subsequently micropropagated to obtain several shoots for field evaluation. Embryo culture has increased the seed germination rate to 9 percent compared to 1 percent germination previously obtained by sowing the seeds directly into nursery potting mixture (Ssebuliba 2000; Ssebuliba et al. 2006).

The first hybrids selected from the crossing of the triploid landraces (3×) and the resistant diploids (2×) were tetraploids (4×, or four sets of chromosomes). They were very fertile and therefore easily set seeds in the presence of a pollen source. The presence of seeds in bananas, especially cooking varieties, poses a quality problem and is not desirable. To circumvent this problem, the tetraploids are crossed with improved diploids to get secondary triploids (3×) that are sterile. As of mid-2005, 11 secondary triploids had been selected for in-depth screening, targeting selection of clones in which the desired traits have been maximally accumulated.

The best-yielding highland bananas proved sterile and hence not amenable to conventional breeding. Genetic engineering, through which desirable genes can be inserted directly into plant cells and thus plants (new genotypes) can be regenerated without passing through seeds, appears to be the only alternative for improving these cultivars. The known method through which new genes can be engineered into bananas is based on cell suspension, where aggregations of single cells are first generated. Desirable trait genes are then inserted into these cells, after which the process of selecting transformed cells and regenerating them into plants for subsequent evaluations follows. Efforts to establish the system are progressing well, and the first plants regenerated from single cells of an EAHB cultivar (Nakyetengu) are planted out in the field to test the extent to which they are true to type (Namanya et al. 2004). Nakyetengu is now in field evaluation, and no somoclonal variation has been observed to date. Three other EAHB cultivars (Mpologoma, Nakinyika, and Nakasabira) have been regenerated and planted for field evaluation. Suspension culture methods are now being optimized for a wide range of cultivars, and arrangements to initiate gene insertions are in their final stages.

Research Introductions of Improved Material

In Uganda, variable response was observed but no useful levels of resistance to weevils, nematodes, and black Sigatoka were reported in endemic materials. As a consequence, nonendemic cultivars that appeared to be resistant were added to the collection. Nonendemic germplasm was acquired mainly via the INIBAP Transit Centre, imported as in vitro plantlets. The plants were evaluated at two stages: (1) on-station and/or selected sites, where the evaluation trials were research controlled, and (2) on-farm multilocational sites, where the evaluation trials were farmers managed with backstopping from extension workers and researchers. The introduced germplasm was planted along with elite cultivars selected from endemic germplasm, which served as checks. At all stages, the germplasm was evaluated for pest and disease response, agronomic performance, and bunch use. Currently, multilocational, on-farm evaluation trials are running in 20 districts of Uganda.

In 1994 and 1999, ARDI received new banana cultivars provided by IPGRI and initiated on-station testing. The cultivars tested included AACV Rose, Cardaba, FHIA01, FHIA02, FHIA03, FHIA17, FHIA23, IC2, Pelipita, Pisang Berlin, Pisang Ceylon, Pisang Sipuru, Saba, SH3436-9, and Yangambi Km 5. The new banana cultivars were thought to have high yield potential and tolerance for one or several combinations of the major banana production constraints.

Multiplication and Dissemination of Planting Material in Uganda

Plant tissue culture was started in Uganda in 1991 by the Banana Based Cropping Systems Research Project in the Department of Crop Science, Faculty of Agriculture, Makerere University, with funding from the Rockefeller Foundation. The first objective was to mass-produce pest- and disease-free planting material as one of the activities to revive banana production in the country. The tissue culture laboratory was constructed at KARI under the Farming Systems Support Project with funding from the E.U. The laboratory that became operational in 1993 was set up to carry out coffee micropropagation on a commercial scale. When activities related to coffee did not use the laboratory to full capacity, research programs working on bananas, potatoes, passion fruits, and vanilla also began to use it. In 1995, banana tissue culture was initiated by the National Banana Research Programme in collaboration with IITA, with the objectives of multiplying (1) disease-free planting materials for research purposes, (2) introduced materials for field evaluation and dissemination, (3) elite cultivars for farmers (mother gardens and direct distribution), and (4) hybrids from international breeding programs (FHIA, IITA). Another objective is in vitro short-term conservation of cultivars and hybrids through periodical subculturing. The laboratories at Kawanda and Makerere University proved unable to provide all the planting materials required by farmers, leading to the construction of the commercial tissue culture laboratory, Agro-genetic Technologies Limited (AGT), in 2002.

Once banana cultivars have passed through on-station or selected-site evaluations, those that perform best with respect to pest and disease response, agronomic characteristics, and end use are selected for multiplication and dissemination. Initial materials from research reach farmers as tissue-culture plantlets. Farmers pass the materials to other farmers in the form of suckers.

The dissemination of farmer-approved planting material involves both the research institute and farmers. At the farm level, various dissemination systems have been reported (Nowakunda et al. 2002). In addition to the informal farmer-to-farmer germplasm exchange discussed above, other farmer distribution systems for banana planting material have been reported in Uganda (Nowakunda et al. 2002).

In one such system, the four farmers hosting the original evaluation banana plots initiated the formation of an association to supply planting materials at a cost. The formation of this association was driven by the belief that when farmers are given free material, they do not attach any value to it or care for the plants. In another instance, an association was formed whereby the farmer who received the evaluation plot was required to pass on 100 free suckers to another farmer, who in turn was required to do the same. Under this system, after a farmer has transferred the initial 100 suckers, he or she is free to sell planting material to anyone, although priority is given to association members.

In a third example, farmers organized themselves into a "community." One of the members of this community was selected to receive and host the evaluation plot, which was to be maintained purely as a "mother garden" belonging to the group. Members were able to obtain free planting material from the garden, which they subsequently multiplied in their own banana groves.

In a fourth instance, farmers were not organized into groups. The original plot was maintained as a mother garden/demonstration plot by the contact farmer with the support and supervision of the agricultural extension program of the subcounty. Extension workers occasionally organized training workshops at the site, and, after attending these workshops, farmers were given free planting materials. Although plants from this garden were not for sale, some farmers who obtained planting material from the mother gardens did sell them or give away free samples.

Of course, farmers who have sufficient resources can purchase tissue-culture plantlets at commercial rates from a private laboratory (AGT) based in Kampala or public laboratories at Kawanda and Makerere University. Kawanda and AGT are currently piloting community-based weaning nurseries of plants produced by tissue culture in central Uganda as another source of planting materials.

Although a rigorous analysis has not been done, banana researchers and extension staff believe that farmer-participatory dissemination systems may be more efficient than other systems for the dissemination of cooking bananas for local consumption, particularly when compared with export-quality dessert bananas, such as those produced in neighboring Kenya for the urban market in Nairobi. One approach to supplying improved planting material is to maintain large planting-stock nurseries in the project area for direct sale to farmers and wholesaling to stockists; another is to establish nurseries managed by "expert" farmers through community organizations.

Multiplication and Dissemination of Planting Material in Tanzania

In 1997, KCDP contracted ARDI to conduct on-farm testing of the introduced banana cultivars. Intensive testing (supervised by researchers) was conducted in Bukoba District and extensive testing (supervised by KCDP staff) was conducted in other districts (Karagwe, Ngara, Biharamulo, and Muleba) of the Kagera Region, where researchers provided support services.

Based on the results of the on-station testing and farmers' assessment, 4 of the 12 cultivars (FHIA01, FHIA02, FHIA03, and Yangambi Km 5) were recommended for further testing on farms with farmers' management. Both intensive and extensive trials were undertaken. Intensive trials involved distribution and testing by 10 farmers per village (30 total) in Kiilima, Kabirizi, and Byamutemba. Extensive trials included three villages per district, and five farmers per district. In 1999, more nonendemic cultivars were added to on-farm testing in both sets of villages.

While on-farm trials were being conducted, the KCDP continued with importa-

Table 4.2 Names of the superior bananas disseminated by the Kagera Community Development Programme

Cultivar	Year multiplied	Exotic/ hybrid	Origin	On-farm tested
1. FHIA01	1998	Hybrid	Honduras	1997
2. FHIA02	1998	Hybrid	Honduras	1997
3. FHIA03	1998	Hybrid	Honduras	1997
4. FHIA17	1998	Hybrid	Honduras	1999
5. FHIA18	2000	Hybrid	Honduras	No
6. FHIA21	2000	Hybrid	Honduras	No
7. FHIA23	2000	Hybrid	Honduras	1999
8. FHIA25	2000	Hybrid	Honduras	No
9. FHIA22	2000	Hybrid	Honduras	No
10. AACV Rose	1998	Hybrid	Honduras	1999
11. Yangambi Km 5	1998	Landrace	DRC	1997
12. Pisang Ceylon	1998	Hybrid	IITA	1999
13. Pisang Berlin	1998	Hybrid	IITA	1999
14. IC2	1998	Hybrid	IITA	1999
15. Pelipita	1998	Hybrid	IITA	1999
16. Saba	1998	Hybrid	IITA	1999
17. Cardaba	1998	Hybrid	IITA	1999
18. Pisang Sipuru	1998	Hybrid	IITA	No
19. Bita3	1998	Hybrid	IITA	No
20. SH3436-9	1998			1999
21. Paka	1998	Hybrid	IITA	No
22. SH3640	2000	Hybrid	IITA	No
23. Kamaramasenge	2000	Exotic	Rwanda	No
24. CRBP	2000	Hybrid	—	No
25. KCDP1	2000	Hybrid	—	No

Notes: DRC is the Democratic Republic of Congo; IITA is the International Institute for Tropical Agriculture; KCDP is the Kagera Community Development Programme. The — signifies not known.

tion, multiplication, and dissemination of planting materials of new banana cultivars. The imported in vitro plants were hardened at Kyakairabwa nursery (mother garden) near Bukoba Town. All in vitro plants were supplied by the INIBAP Transit Centre, based at the Catholic University of Leuven, Belgium. By July 2002, about 71,000 in vitro plants from 25 different banana cultivars had been imported in Kagera Region (Weerdt 2003). The 25 banana cultivars are shown in Table 4.2.

Some of the banana cultivars proved adaptable to the Bukoba banana/coffee

Table 4.3 Estimated quantities of planting material (numbers of suckers) supplied to farmers during the Kagera Community Development Programme

Year	Direct diffusion	Indirect diffusion	Total number of suckers
1997	1,352	0	1,352
1998	2,330	2,684	5,014
1999	20,285	12,712	32,997
2000	101,420	78,706	180,126
2001	151,588	438,878	590,466
2002	150,333	390,797	541,130
Total	427,308	923,777	1,351,085

Source: KCDP (2002).

Table 4.4 Percentage of survey farmers obtaining information about banana management practices, by source of information

Information source	Percentage of farmers
Formal sources	61.26
Nongovernmental organizations	22.91
Government extension	42.76
Researchers	9.89
Other farmers	90.49
Mass media	25.70
Radio	24.03
Publications	2.83

Source: Katungi (2006).

Note: Percentages do not total 100 because farmers receive information from more than one source.

farming system zones and acceptable to specific communities or categories of households. These include FHIA01, Yangambi Km 5, Pelipita, SH3436-9, FHIA17, and FHIA23. Further monitoring on their field performance was recommended for these cultivars, however. Other cultivars (for example, FHIA02, FHIA03, Cardaba, Pisang Berlin, Pisang Ceylon, and AACV Rose) proved less appealing, because they were not tolerant of local pests and diseases. Further multiplication and dissemination of these cultivars was not recommended (Nkuba, Ndege, and Mkulila 2002).

KCDP involved nongovernmental organizations (NGOs), primary schools, district departments of agriculture, and some individual progressive farmers in the establishment of nurseries and multiplication of superior bananas. Because KCDP supplied farmyard manure and paid laborers, multiplication of these bananas was successful and proceeded rapidly.

KCDP began to distribute planting material of superior bananas to farmers by the end of 1997 free of charge, later charging only 100 Tsh per sucker. Among the people of this region, banana suckers are traditionally exchanged free of charge, and thus, few farmers were willing to pay for them. According to the project report, an estimated 1 million banana suckers had diffused among farmers in Kagera Region by the year 2002 (Table 4.3). These high numbers were achieved through direct (nursery-to-farmer) and indirect (farmer-to-farmer) diffusion (Gallez et al. 2002). In direct diffusion, farmers obtain new planting materials from multiplication, trials, or demonstrations. The numbers reported in Table 4.3 do not reflect observed transfers to or among farmers, but are estimated numbers of suckers available for planting based on a rate of two to three suckers per mat planted in the previous year.

Dissemination of Management Practices in Uganda

Three types of mechanisms have been used in the dissemination of banana cultivars and management practices recommended by NARO (Tushemereirwe et al. 2003) in Ugandan villages: (1) formal sources (the extension services—government, NGOs, and on-farm research), (2) other farmers, and (3) mass media. Farmers surveyed by Katungi (2006) were asked about their sources of information for the practices they know about. Responses are summarized in Table 4.4.

The most widely used mechanism for information about recommended management practices is farmer-to-farmer dissemination, in all the production regions. More than 90 percent of farmers surveyed reported that they obtained information regarding banana management from other farmers, and nearly two-thirds (61.26%) were provided information by formal sources. Farmers often received information from multiple sources. Of formal sources, government extension was more frequently reported perhaps due to the wider coverage and the historical involvement in dissemination. The dissemination of new agricultural technologies in Uganda was traditionally the role of the government extension service, which was joined by some NGOs in the 1990s as part of an economic recovery program ushered in by President Museveni's government.

On-farm research and mass media are relatively new mechanisms introduced especially in the Central Region as a strategy to revive banana productivity. As expected, researchers are popular as a source of information in this region, because on-farm research intervention has been more concentrated there compared to other regions. A quarter of farmers reported receiving information through mass media. Radios are a frequent source of information in some areas, but publications are rarely used.

Conclusions

Pest and disease problems have long beset banana production in Uganda and Tanzania. To address these problems, banana researchers in the Lake Victoria region have sought host-plant resistance through assembly and evaluation of endemic and nonendemic germplasm as well as crossing and genetic transformation. They have also recommended management practices to reduce pest and disease damage, and support soil fertility. Farmers have continued to search for fresh planting material nearby and at great distances.

Distinct approaches to disseminating new materials were pursued in Uganda and Tanzania. In Tanzania, the immediate impact of KCDP was impressive in terms of numbers of plantlets distributed, but the longer-term effects will need to be observed with the passage of time. Chapter 9 provides some additional information regarding the impact of KCDP.

NARO in Uganda holds the perspective that farmer-participatory selection and dissemination improves the confidence of farmers in new banana cultivars, helping them to acquire skills and develop criteria to select genotypes suitable for them. Given the reproductive properties of bananas, planting-material distribution systems designed by farmers themselves appear to be most appropriate, enabling step-by-step and systematic diffusion of improved cultivars to farming communities. Nonetheless, closer inspection of alternative dissemination models and mechanisms would be necessary to verify which are more cost effective or efficient, and why.

As in the case of planting material, the most common sources of information about recommended banana management practices is other farmers, although extension services and researchers have been popular sources of information in some localities where public efforts have been focused. Radio is also an important source of information, but use of publications by farmers is negligible.

References

Bosch, C. H., A. Lorkeers, M. R. Ndile, S. R. B. Mgenzi, and E. Sentozi. 1996. *Diagnostic survey: Constraints to banana productivity in Bukoba and Muleba districts, Kagera Region, Tanzania.* Working Paper 8. Bukoba, Tanzania, and Amsterdam: Agricultural Research Institute Maruku and Royal Tropical Institute.

FSR (Farming Systems Research). 1989. *Diagnostic study of Bukoba District, part II: The formal study.* Working Paper 5. Bukoba, Tanzania, and Amsterdam: Maruku Agricultural Research Institute and Royal Tropical Institute.

Gallez, A., O. Machiels, C. Mbehoma, R. Mitti, and G. Patrick. 2002. *Impact of superior banana cultivars (SBVs) in Kagera, Tanzania.* KCDP annual report. Bukoba, Tanzania: Kagera Community Development Programme.

Gold, C. S. 1998. Banana weevil: Ecology pest status and prospects for integrated control with emphasis on East Africa. In *Proceedings of a symposium on biological control in tropical habitats: Third International Conference on Tropical Entomology,* ed. S. K. Saini. Nairobi, Kenya: International Centre of Insect Physiology and Ecology.

———. 2000. Biology and integrated pest management of banana weevil, *Cosmopolites sordidus* (Germar). In *Advancing banana and plantain R&D in Asia and the Pacific,* Vol. 10, ed. A. B. Molina, V. N. Roa, and M. A. G. Maghuyop. Montpellier, France: International Network for the Improvement of Banana and Plantains.

Gold, C. S., G. Night, A. Abera, and P. R. Speijer. 1998. Hot-water treatment for the control of the banana weevil, *Cosmopolites sordidus* Germar (Coleoptera: Curculionidae), in Uganda. *African Entomology* 6 (2): 215–221.

Gold, C. S., N. D. T. R. Rukazambuga, E. B. Karamura, P. Nemeye, and G. Night. 1999. Recent advances in banana weevil biology, population dynamics and pest status with emphasis on East Africa. In *Mobilizing IPM for sustainable banana production in Africa. Proceedings of a workshop on banana IPM,* Nelspruit, South Africa, 23–28 November 1998, ed. E. A. Frison, C. S. Gold, E. B. Karamura, and R. A. Sikora. Montpellier, France: International Network for the Improvement of Banana and Plantains.

Gold, C. S., J. E. Pena, and E. B. Karamura. 2001. Biology and integrated pest management for the banana weevil *Cosmopolites sordidus* (Germar) (Coleoptera: Curculionidae). *Integrated Pest Management Reviews* 6 (2): 79–155.

Kabwoto, E. 1974. Buhaya agricultural systems: Colonial disruption and present problems: A research note. *Journal of the Geographical Association of Tanzania* 10: 66–79.

Katungi, E. 2006. Social capital and technology adoption on small farms. A case of banana production technology in Uganda. Ph.D. dissertation, University of Pretoria, Pretoria, Republic of South Africa.

KCDP (Kagera Community Development Programme). 2002. *Annual report.* Bukoba, Tanzania: KCDP.

Mbwana, A. S. S., M. S. R. Byabachwezi, and S. I. Mkulila. 1996. *Banana germplasm collection.* Bukoba, Tanzania: Agricultural Research Institute, Maruku.

Murashige, T., and F. Skoog. 1962. A revised medium for rapid growth and bio-assays with tobacco tissue cultures. *Physiolologia Plantarum* 15: 473–497.

Namanya, P., S. M. Magambo, G. Mutumba, and W. K. Tushemereirwe. 2004. Somatic embryogenesis from immature male inflorescence of East African highland banana cv "Nakyetengu." *African Crop Science Journal* 12 (1): 47–53.

Nkuba, J. M., L. J. Ndege, and S. Mkulila. 2002. *On-farm testing and acceptability of improved banana cultivars in Bukoba District, Kagera Region (1997–2001).* Field Note 144. Bukoba, Tanzania: Agricultural Research and Development Institute, Lakeshore-Maruku.

Nowakunda, K., W. K. Tushemereirwe, C. Nankinga, P. Nahamya, and P. Ragama. 2002. Evaluation and dissemination of introduced banana germplasm: A preliminary assessment of the planting material distribution systems. Paper presented at the Rockefeller Foundation/National Agricultural Research Organization meeting on biotechnology, breeding and seed systems for African crops, November 4–7, Entebbe, Uganda.

Ortíz, R., and D. Vuylsteke. 1994. Plantain breeding at IITA. In *The improvement and testing of Musa: A global partnership,* ed. R. Jones. Montpellier, France: International Network for the Improvement of Banana and Plantains.

Rald, J., and K. Rald. 1975. *Rural organisation in Bukoba District, Tanzania.* Uppsala, Sweden: The Scandinavian Institute of African Studies.

Ssebuliba, R. N. 2000. Fertility in East African highland bananas. Ph.D. thesis, Makerere University, Makerere, Uganda.

Ssebuliba, R. N., D. Talengera, D. Makumbi, P. Namanya, A.Tenkouano, W. K. Tushemereirwe, and M. Pillay. 2006. Reproductive efficiency and breeding potential of East African highland bananas. *Field Crops Research* 95 (2–3): 250–255.

Tushemereirwe, W. K., A. Kangire, J. Kubiriba, and K. Nowakunda. 2000. Acceptability of *Fusarium* wilt resistant banana cultivars considered appropriate replacements for a cultivar susceptible to one disease in Uganda. *Uganda Journal of Agricultural Sciences* 5 (6): 62–64.

Tushemereirwe, W. K., I. Kashaija, W. Tinzaara, C. Nankinga, and S. New. 2003. *A Guide to Successful Banana Production in Uganda.* Second edition. Entebbe, Uganda: National Agricultural Research Organization.

Weedrt, J. 2003. *Adoption of superior banana cultivars in the Kagera Region: Accomplishment and constraints.* Bukoba, Tanzania: Kagera Community Development Programme and Economic Development Initiatives, Limited.

CHAPTER 5

Characteristics of Banana-Growing Households and Banana Cultivars in Uganda and Tanzania

Svetlana Edmeades, Melinda Smale, Enoch Mutebi Kikulwe, Jackson M. Nkuba, and Mgenzi Said Ramadhan Byabachwezi

A first stage of data analysis confirmed few statistical differences between exposed and nonexposed areas in the underlying social and demographic characteristics of households and the physical characteristics of their farms. This result validates the sample design (Chapter 2). In this chapter, selected characteristics of banana-growing households in Uganda and Tanzania are analyzed by elevation, and uses of new banana cultivars, practices, and planting material systems are summarized by exposure and elevation.[1] In general, means, frequencies, or proportions only are reported here, weighted by the inverse probability of selection.

Banana-Growing Households

Household demographic characteristics in the Lake Victoria region of Uganda and Tanzania are shown in Table 5.1. Household heads averaged 45 years of age with roughly 6 years of education in Uganda, and 49 years of age with 7 years of education in Tanzania. Though the age of the household head and mean years in school are statistically different between the highland and lowland areas of Uganda, the magnitudes of the differences are not meaningful. Roughly a quarter of household heads are reported to be women, with a smaller percentage in high-elevation areas in Uganda, and a comparable pattern in Tanzania. In Tanzania, gender of the household head is the only major demographic characteristic that differs statistically by elevation. Clearly, longer-term absence of men is more prevalent in the lowland areas.

There are a total of 777 households in the sample data analyzed in this chapter, of which 517 are in Uganda and the remaining 260 are in Tanzania.[2] Given high population densities in the lowlands of Uganda, the majority of households (419) are located in the low-elevation areas,

[1] Additional statistical summaries and further details can be found in Edmeades et al. (2004), Katungi (2006), Bagamba (2007), and Nkuba (2007). Statistics on village social structure are found in Appendix E.

[2] The original sample size comprised 800 households across the banana-production domain in Uganda and Tanzania. However, despite location in the major banana-producing areas of Uganda, 23 households in the Ugandan subsample did not grow bananas and were excluded from the analysis.

Table 5.1 Social and demographic characteristics of households, by elevation, Tanzania and Uganda

	Tanzania			Uganda		
	Elevation			Elevation		
Characteristic	Low	High	All	Low	High	All
	Mean					
Age of household head (years)	48.6	48.3	48.5	45.2**	41.9**	44.9
Education of household head (years)	6.4	6.7	6.5	5.7*	4.9*	5.8
Household size (number of individuals)	6.2	7.9	6.6	5.7	5.4	5.6
	Percentage					
Dependency ratio	24.7	33.7	26.9	49.0	50.6	49.1
Households with members residing away	36.1	18.6	30.9	16.6	19.0	16.8
Female household head	26.4***	17.5***	25.0	25.2***	18.4***	24.4

Notes: ***, **, and * indicate statistical significance at the 1, 5, and 10 percent levels, respectively, in the difference of means or distributions across elevations within countries. Dependency is defined as the percentage of household members younger than 15 years or older than 64 years.

with 98 households residing in high-elevation areas. Similarly, most of the households in Uganda (355) are in exposed areas, and only 162 are in nonexposed areas. In Tanzania, 220 households are located in low-elevation areas and 40 are in high-elevation areas. In contrast to Uganda, 220 households in Tanzania are located in exposed areas and 40 are found in areas that have not been exposed to formal programs to introduce banana cultivars (see Appendix D for details regarding the sample survey design).

The size of households in Uganda is, on average, five persons, with dependents (children younger than 15 years and adults older than 64 years) constituting about half of the household. Household sizes average six persons in Tanzania, although the dependency ratio tends to be nearly half the level observed in Uganda. Nearly a third of households surveyed in Tanzania reported members residing away from home, with a much higher percentage in the lowlands. The overall percentage of households with members residing away from home appears to be lower in Uganda than in Tanzania, with no difference by elevation.

Assets of household farms are composed of (1) the value of crops in storage, livestock, other consumer durables, such as radios, bicycles, and motorcycles; (2) financial assets, or credit; and (3) farm land (Tables 5.2 and 5.3). In Uganda, crops in storage at the time of the survey included, in decreasing order of total value (if sold at that time) groundnuts, beans, maize, millet, rice, cassava, soybeans, coffee, sweet potatoes, sorghum, peas, and tobacco. More millet and less cassava and maize were stored by households of high-elevation areas, corresponding to millet's relative importance in their diet. Millet was the major staple in the southwestern highlands before bananas; maize and cassava are the primary staples in the eastern areas of the country, found in the low-elevation stratum of the sample domain. Sweet potatoes were found only in the eastern and southwestern parts of the country, and are a major crop in the Eastern Region.

Table 5.2 Average total value per household of crops stored, livestock, and other assets, and percentage distribution by type of asset, Uganda (Ush)

	Elevation		
Asset	Low	High	All
Crops in storage	23,730	21,663	23,508
Livestock	326,934	264,985	320,287
Other assets	49,957	72,325	52,357
Percentage distribution			
Crops in storage	19.9***	28.5***	20.8
Livestock	49.3***	32.9***	47.6
Other assets	30.8**	38.6**	31.6
Total	100	100	100

Notes: *** and ** indicate statistical significance at the 1 percent and 5 percent levels, respectively, in the difference of means or distributions across elevations within countries. Other assets include radios, bicycles, and motorcycles.

Table 5.3 Average total value per household of crops stored, livestock, and other assets, and percentage distribution by type of asset, Tanzania (Tsh)

	Elevation		
Asset	Low	High	All
Crops in storage	32,765***	38,234***	34,400
Livestock	176,559***	330,695***	222,621
Other assets	135,494***	171,159***	146,152
Percentage distribution			
Crops in storage	18.7**	11.1**	16.9
Livestock	29.3***	41.1***	32.2
Other assets	52.0**	47.7**	50.9
Total	100	100	100

Notes: *** and ** indicate statistical significance at the 1 percent and 5 percent levels, respectively, in the difference of means or distributions across elevations within countries. Other assets include furniture, radios, bicycles, and motorcycles.

In Tanzania, the most important food crops kept in storage at the time of the survey included beans, maize, groundnuts, bambara nuts, and sorghum. Other crops found stored by farmers were dried cassava, soybeans, cowpeas, and coffee. High-elevation areas have higher average total values of crops in storage than do low-elevation areas.

In Uganda, livestock include cattle, chickens, goats, sheep, and pigs. Livestock contributes the highest overall value (49 percent) of the three asset categories for which value was easily assessed with market prices. Although levels did not differ significantly because of underlying variation, proportional distributions among asset categories clearly depend on elevation, with the value of stored crops and consumer durables being greater in the highlands. Because cattle were distributed across much of the area in either elevation zone, and chickens were concentrated among villages in the low-elevation areas, livestock assets contribute proportionately more to total assets in the low-elevation zone. In part because of wide variation among households, differences in total value of livestock are not statistically significant—except in the case of chickens. In high-elevation areas, chickens are less important in the diet. They also tend to destroy mulch in banana groves, imposing a constraint on banana production. Goats also appear to be relatively more important in high-elevation areas.

In Tanzania, households keep cattle, goats, sheep, pigs, chickens, and ducks, and livestock represents the largest in value on average of the three major asset categories (crops stored, livestock, and other consumer durables). The number of farmers keeping livestock types and average number of each type per household differ statistically across the agroecological zones, as do asset values by elevation. Households in the highlands have a higher value of each type of asset than those in the lowlands. Livestock value (Tsh 330,695) in the highlands was about twice that in the lowlands (Tsh 176,559). Percentage distributions by type of asset

Table 5.4 Percentage of farmers using credit, and use of credit, by elevation, Uganda

	Elevation		
	Low	High	All
Households that sought credit	13.2**	8.9**	12.8
Households that obtained credit, of those who sought it	49.6	88.3	52.5
Use of credit			
Agricultural	35.3	43.0	36.2
Basic needs	50.0	57.0	50.8
Trade	14.7	0.0	13.0
Total	100.0	100.0	100.0

Note: ** indicates statistical significance at the 5 percent level in the difference of means or distributions across elevations within countries.

differ somewhat, although in each elevation, other assets represent around half of the total asset value.

Almost two-thirds of Ugandan households (65 percent) had radios, bicycles, and motorcycles, with an average value of Ush 52,357 and similar to other assets, wide variation in total values across households. Motorcycles were more common in high-elevation areas. At least 50 percent of the Tanzanian households owned basic tools for cultivation (hand hoes, pangas, sickles, forked hoes, axes, and gunny bags), furniture (tables, chairs, or beds), a radio, and a bicycle. Three households had no hand hoes. Less than a quarter of households had wheelbarrows, pruning knives, bow saws, or knapsack sprayers.

Very few farmers in the sample area of Uganda sought credit, with households in the low-elevation areas more likely to seek it (Table 5.4). Of those who did, however, a much higher percentage actually obtained it in the high-elevation areas—although the numbers of credit seekers are too small to permit generalization. The few households using credit directed it toward basic necessities, such as food, health, and school fees, followed by hiring labor. Buying inputs for agriculture-related activities was least important as an expenditure category, and its direct role in the adoption of new banana technologies is therefore thought to be negligible. The main sources of credit were individual moneylenders and NGOs.

There was no recorded use of formal credit in the Tanzanian sample. Farmers in Tanzania generally obtained informal loans or credit to spend on buying food and paying medical charges, school fees, and other home expenses. Less than 2 percent used credit in agriculture, and only a total of 13 percent used credit at all. Major sources of credit were friendship (61 percent), moneylenders (21 percent), farmer groups (10 percent), and NGOs (7 percent).

Farm sizes are on average 1 ha larger in the high-elevation areas of Tanzania compared to the lowland areas, although cropped land areas are not statistically different in size (Table 5.5). Households in low-elevation areas of Uganda appear to own larger portions of land than those in high-elevation areas, although the mean cropped areas do not differ significantly, and they appear to be smaller than in Tanzania. Differences in the size of farms between the low- and high-elevation areas in Uganda are often attributed to population pressures that have led to more pronounced fragmentation of land in the highlands (Nkonya et al. 2005). Slightly over a third of total cropped land is allo-

Table 5.5 Average land owned and cropped, banana as a percentage of total cropped area, and age of banana plantations, by elevation, Tanzania and Uganda

	Tanzania			Uganda		
	Elevation			Elevation		
Characteristic	Low	High	All	Low	High	All
Total land owned (ha)	1.6	2.7***	1.9	1.5**	0.9**	1.4
Cropped land (ha)	1.0	1.9	1.2	0.7	0.6	0.7
Banana as percentage of total cropped area	38.1	35.0	35.7	19.7***	39.3***	22.1
Age of banana plantation (years)	37	27	36	10***	29***	12

Note: *** and ** indicate statistical significance at the 1 percent and 5 percent levels, respectively, in the difference of means or distributions across elevations within countries.

Table 5.6 Percentage of households participating in banana markets, selling or buying bananas, by elevation, Uganda and Tanzania

	Uganda			Tanzania		
	Elevation			Elevation		
Type of participation	Low	High	All	Low	High	All
No Participation	22.55	24.07	22.72	41.36***	10.00***	36.54
Only sells	31.92***	62.82***	35.38	44.09***	90.00***	51.15
Only buys	32.03***	5.21***	29.02	10.00	—	8.46
Sells and buys	13.50*	7.90*	12.87	4.55	—	3.85

Notes: Differences across elevations within countries are tested by cross-tabulation using the chi-square distribution. *** and * indicate statistical significance at the 1 percent and 10 percent levels, respectively; — signifies no households in this category.

cated to bananas in both the highlands and lowlands in Kagera Region, with no apparent difference by elevation. In Uganda, farmers in the highlands allocate nearly twice as great a share of cropped area to bananas (39 percent) as those in the lowlands (20 percent), close to the levels observed in Tanzania. Banana growers in the high-elevation areas of Uganda have tended their banana plantations for considerably longer than those farming in the lowlands, probably because pest and disease pressures are lower. These farmers are also driven by commercial objectives in their production decisionmaking rather than by a need to satisfy the food requirements of their families. In Tanzania, the difference in the age of plantations is not significant between zones, and the overall average age is 36 years.

Accordingly, household participation in banana market transactions differs significantly by elevation. Some households choose not to participate, while others participate only as sellers, only as buyers, or as both sellers and buyers (Table 5.6). The majority of households in the survey domain report some involvement in banana markets, although roughly a quarter of the households remain autarkic (without any market-based purchase or sales of bananas). In high-

Table 5.7 Percentage of banana production, sales, and use decisions made by women, by elevation, Uganda and Tanzania

	Uganda			Tanzania		
	Elevation			Elevation		
Action	Low	High	All	Low	High	All
"Own" the plot	36.4***	17.0***	34.2	9.6***	0.0***	8.1
Choose which banana cultivars to grow	53.6***	17.0***	49.5	12.9***	0.0***	9.8
Make final production decisions	47.5***	18.6***	44.3	14.9***	0.0***	11.3
Sell bunches of beer bananas	48.1*	32.1*	46.4	23.2***	7.9***	19.2
Sell bunches of cooking bananas	45.5*	34.9*	43.6	37.0***	67.6***	46.3
Sell bunches of other banana types	49.4*	31.3*	47.8	36.1***	75.0***	48.5
Cook matooke	90.3	88.9	90.1	79.8	80.8	80.1
Make beer/juice	48.1***	4.3***	44.9	36.5***	8.4***	27.9

Notes: *** and * indicate statistical significance at the 1 percent and 10 percent levels, respectively, in the difference between distributions across elevations within countries. The data are based on decisions made for the major banana plot only.

elevation areas, given the commercial nature of banana production, market participation is mostly associated with selling of banana bunches at the farm gate. Buying banana bunches is a more common practice among households in low-elevation areas than it is in high-elevation areas, reflecting the geographical shift in the locus of banana production from the Central Region to the southwestern highlands, from where it is transported and distributed across markets in the lowlands.

Banana farmers in the Tanzanian highlands are twice as likely to sell bananas than farmers in the lowlands, none of them both sell and buy, and none buy only (Table 5.6). Forty-one percent of farmers in the low-elevation areas choose not to participate in banana markets at all, compared to only 10 percent in the high-elevation areas. The commercial orientation of highlands farmers is evident, as also tends to be the case in Uganda.

Findings confirm anecdotal evidence that women in Uganda are more heavily involved in banana production, sale, and use in the lowlands than in the highlands (Table 5.7). Data were collected about decision-making by plot and are shown for the major banana plot in the table. In low-elevation areas, over half of those who make final decisions on banana cultivar choice are women, and nearly half make final decisions about banana production, sales, and purchase, although only slightly more than one-third owned the plot. The lowland areas of central and eastern Uganda also represent the historic locus of production of bananas, where tradition and experience give greater and broader responsibility to women. The lesser involvement of women in banana-related activities in high-elevation areas could reflect the commercial nature of banana production in southwest Uganda. Men appear to be in charge of the whole production chain, from owning the plot to deciding on the cultivars grown, the management practices applied, and sales of different types of bunches at the farm gate, in nearby markets, or distribution centers. Regardless of production zone, women are responsible for preparing the food. In the higher-elevation areas, beverage making from banana is evidently the responsibility of men, while nearly

Table 5.8 Average number of banana cultivars per household and per village, Uganda

	Elevation		
	Low	High	All
Banana cultivars per household	6.73***	9.07***	6.99
Cooking cultivars per household	4.42***	5.57*	4.55
Banana cultivars per village	22.91	24.80	23.33

Note: *** and * indicate statistical significance at the 1 percent and 10 percent levels, respectively, in the difference of means or distributions across elevations within countries.

half of those who make beverages from bananas on the plots in the low-elevation areas are women. Sales of banana beverages can be an important income-generating activity.

In Tanzania, women who own plots have usually inherited the banana plots from their husbands or from their parents. When they are elderly, the ownership and control of the banana plots is transferred to sons and very rarely to daughters. Of all banana plots recorded during the household survey, about 92 percent were declared to be owned by men (mainly heads of households and in rare cases, fathers of heads of households). The majority of men also reported that they made cultivar choices, compared to a minority of women. This pattern is much more pronounced in the commercially oriented highland areas.

Overall, men are more likely to sell bunches of brewing bananas, as well as to make juice or beer, because of their importance in generating cash. At the time of the survey, we found that brewing bananas were primarily sold by men in local markets. Up until the 1980s, when household cash income was largely generated through coffee sales, decisions about selling bananas were most often made by women. Women's participation in sales of cooking bananas is still substantial. The increased market importance of bananas seems to have resulted in greater involvement of men in sales, however, and particularly in transactions with brewing types. There is more involvement of women in sales of beer types and in making beverages in the lowlands.

Banana Cultivars

Survey data confirm the high level of banana cultivar diversity both in the aggregate (the country) and at the microlevel (on single farms) in both Uganda and Tanzania. In Uganda, 95 banana cultivars are currently grown only by sample farmers. The majority of the cultivars (86 percent) are endemic to East Africa (AAA-EA genomic group). The remaining 14 percent are composed of exotic cultivars or banana hybrids (both classified here as nonendemic) introduced into East Africa (see further descriptions in Chapter 2 and Appendix A).

An average of 23 different banana cultivars was identified at the village level across the sampled villages in Uganda (Table 5.8). Surveyed households grow a large number of different banana cultivars simultaneously on their farms. Farmers tend to grow two or three different use groups. On average, households grow more cultivars (in general, and in particular, of cooking types) in high-elevation areas. This finding could be associated with the larger scale of banana production in high-elevation areas.

Major cultivars appear to be fairly uniformly distributed across households. That is, the cultivars most frequently grown by farmers (percentage of households) on their

Table 5.9 Most common banana cultivar sold, prices, and time to market, by elevation, Uganda

	Elevation		
	Low	High	All
Households participating as sellers			
Most common cultivar sold (percent of all cultivars sold)	Nakitembe (10.52)	Mbwazirume (8.74)	Nakitembe (9.54)
Average farm-gate price (Ush)	798.26	863.93	803.75
Households participating as buyers			
Time taken to reach point of purchase (hours)	1.12***	0.65***	1.07
Average market price (Ush)	1,643.49	1,874.78	1,649.11

Note: *** indicates statistical significance at the 1 percent level in the difference of means or distributions across elevations within countries.

principal banana plots are generally the same as those most widely planted (percentage of mats). Among them, the endemic cooking bananas predominate. Even the most popular banana cultivars occupy less than 10 percent of all mats grown by all farmers surveyed, however, reflecting the high diversity of banana cultivars found within and among farms in Uganda.

Cooking types are the most frequently sold and purchased banana cultivars at the market. Thus, cooking bananas, a food staple, are important generators of cash in local markets. No significant differences in average farm-gate prices and average market prices are observable among major cultivars sold or purchased across elevation (Table 5.9). Households in the lowlands take longer to reach the point of purchase. Bicycles are the most frequent means of transportation of banana bunches to/from point of sale/purchase.

A total of 120 banana cultivars were identified in the sample of Tanzanian farmers, of which 96 (80 percent) are endemic, 9 exotic, and 8 hybrids (see Appendix B). Based on common use, 79 banana cultivars were classified as cooking bananas, 20 were brewing bananas, 4 were dessert bananas, and 3 were roasting types. Banana cultivars with multiple uses were also recorded. As in Uganda, the most commonly grown group of cultivars across elevations and subregions are the endemic cooking types. In contrast to Uganda, a few exotics, such Yangambi Km 5 and the hybrid FHIA17, were also among the 20 most frequently grown cultivars in the lowlands.

The number of banana cultivars per household in Tanzania ranges from 3 to 29, with an overall average of 10 cultivars per household. All except two farmers were found growing at least one of the endemic cultivars (Table 5.10). The number of endemic cultivars grown per household ranges from 1 to 16, with an average of 6. The number of exotic cultivars ranges from 1 to 13, with an average of 4 per banana grove. Up to four hybrids are grown, averaging less than one per grove in the highlands.

Banana cultivars are bundles of attributes related to production traits and the consumption goods the harvest produces (see Chapters 2 and 6). Each cultivar supplies a combination of attribute levels, and demand for attributes is specific to each farmer, varying with farmer characteristics. Previous field research in Uganda was used to identify a set of cultivar attributes (related to phenotypic characteristics, which in turn reflect genotype, at least in part) that farmers consider when choosing cultivars. Farmers were asked to rate cultivars according to the attributes supplied in a scale of 3 (very

Table 5.10 Average number of banana cultivars per household, by genomic group, Tanzania

	Elevation		
Genomic group	Low	High	All
Endemic	6.5	6.6	6.5
Nonendemic			
Exotic	4.0	2.6	3.7
Hybrids	0.4	0.02***	0.3
All cultivars	10.8	9.2	10.4

Note: *** indicates statistical significance at the 1 percent level in the difference of means or distributions across elevations within countries.

Table 5.11 Average rating of endemic and hybrid bananas, by attribute, Uganda and Tanzania

	Uganda		Tanzania	
Trait	Endemic	Hybrid	Endemic	Hybrid
Cooking quality	2.9**	2.0**	2.6**	2.1**
Suitability for beer	1.1**	1.6**	1.3**	1.7**
Bunch size (yield)	2.2**	2.9**	2.0**	2.3**
Resistance to black Sigatoka	2.0	2.4	2.3**	2.7**
Resistance to *Fusarium* wilt	2.5	2.7	3.0**	2.8**
Resistance to weevils	1.5**	2.4**	1.7**	2.6**

Notes: ** indicates statistical significance at the 5 percent level. In Uganda, 23 households grow both hybrid and endemic banana cultivars in their groves. In Tanzania, 50 households grow both groups of cultivars simultaneously. Rating system: 1 = bad, 2 = neither good nor bad; 3 = very good.

good), 2 (neither good nor bad), and 1 (bad). Average ratings of hybrid and endemic cultivars are compared for the households in the sample growing both types of cultivars in Table 5.11.

In both Uganda and Tanzania, findings support banana researchers' observations that farmers perceive endemic bananas to be superior in terms of consumption quality characteristics. Hybrid cultivars appear to be better suited for beer making than endemic types, because they are used for multiple purposes in both Uganda and Tanzania. Bred for favorable production traits, hybrids are found to yield larger bunches in Tanzania—but not among Ugandan growers. Tanzanian farmers perceived hybrids as more resistant to black Sigatoka and, in both Uganda and Tanzania, weevils. Endemic cultivars were perceived to be more resistant to *Fusarium* wilt in Tanzania, as is expected given the genomic group.[3] Though the data

[3] The score for *Fusarium* wilt among endemic cultivars clearly supports the view established in the banana literature that farmers often confuse cause and effect (Gold et al. 1998, 2004). In our sample, we found strong correlation between the attributes "resistance to *Fusarium* wilt" and "resistance to weevils," which is attributed to farmers' misperceptions of cause, because the effect of both biotic constraints is visually similar on the plant.

Table 5.12 Results of hypothesis tests comparing distributions of expected yield loss due to three constraints, by elevation, across use groups, Uganda

	Kolmogorov-Smirnov test (p-value)		
Use group	**Black Sigatoka**	**Weevils**	***Fusarium* wilt**
Cooking cultivars	0.000	0.000	0.764
Beer cultivars	0.000	0.931	0.000
Sweet cultivars	0.000	0.324	0.412
Multiuse cultivars	0.359	0.775	0.099
Roasting cultivars	0.049	0.010	0.928

do not confirm the same relationship for Uganda, it should be borne in mind that the number of farmers growing hybrids, and thus the number of observations compared, are few. A t-test comparing the expected yield losses for endemic and hybrid bananas, for the farmers in Tanzania who grow both, indicates that they are on average less susceptible at a 10 percent significance level. These data, though sparse, provide some evidence that farmers perceive benefits from the introduction of hybrids in Tanzania. The evidence is less clear in Uganda, perhaps as a consequence of the smaller volume of hybrids disseminated, differences in disease pressures, and farmers' knowledge of the diseases.

In Uganda, expected yield loss distributions were compared for the three biotic constraints of interest (black Sigatoka, *Fusarium* wilt, and weevils) by elevation across use types. The p-values of test results comparing the distributions are shown in Table 5.12.[4] Data support the hypothesis that the underlying statistical distributions of expected yield losses from black Sigatoka differ by elevation for cooking, beer, sweet, and roasting cultivars. The geographical focus of the disease is the low-elevation areas, with limited incidence in the highlands. This pattern is less evident for *Fusarium* wilt and weevils, with no statistically significant differences being evident between expected yield loss distributions for most banana use groups. The importance of altitude in the incidence and severity of biotic pressures that farmers perceive supports biophysical studies conducted previously and the sample stratification scheme (Gold, Kiggundu, and Abera 1998; Gold et al. 2004), although the data indicate that leaf spots and weevils are less important considerably above 1,200 m.a.s.l., the dividing line used in this study.

Use of Recommended Technology

In Uganda, an introduced landrace, hybrids, and elite endemic bananas have been diffused to farmers. In each community where these have been introduced, endemic

[4] Expected yield losses were calculated in the following manner. First, farmers were asked about the incidence of the biotic pressure (in years) they had experienced during their involvement in banana production. Next, enumerators elicited triangular yield distributions (minimum, maximum, and mode) in the presence and absence of the biotic pressures from farmers. Physical differences in the effects of pathogens on different banana cultivars were readily observable to them, and photos were provided to assist them in recognition. Expected yield losses were calculated, per biotic constraint and cultivar, from the elicited yield distributions (Hardaker, Huirne, and Anderson 1997; see Chapter 10 for formulas).

Table 5.13 Percentage of farmers growing improved banana cultivars, Uganda

	Elevation		Exposure		
Cultivar type	Low	High	Exposed	Not exposed	All
Exotic or hybrid	2.5**	0.0**	3.6	2.0	2.2
Elite endemic	7.5**	0.0**	6.2	6.7	6.7
All improved or introduced	9.9**	0.0**	9.5	8.7	8.8

Notes: ** indicates statistical significance at the 5 percent level in the difference of means or distributions across elevations within the country. Exotic refers to introduced farmers' cultivars; elite endemic refers to endemic types selected and enhanced by the National Agricultural Research Organization; and all improved or introduced combines exotic, elite endemic, and hybrid categories.

EAHBs continue to dominate banana groves. Yangambi Km 5 is a landrace or farmer cultivar, grouped here with hybrids only because it has been introduced recently. Hybrids include FHIA01, FHIA03, FHIA17, and FHIA23. The elite endemic cultivar is Mpologoma. Mpologoma was selected, propagated with tissue culture, and disseminated by NARO. The estimated fraction of farmers using any type of improved banana in Uganda is only 9 percent (Table 5.13), all of whom are located in the low-elevation areas. Farmers in commercial growing areas of southwestern Uganda are generally more content with their available clones than those in less productive areas, such as the lowland areas of central Uganda, where productivity has declined.

The average share of the banana groves allocated to improved cultivars among users is 13 percent. The numbers and mat shares in the exposed areas are both higher for elite endemic types than for exotics or hybrids (Table 5.14). Elite endemic types are preferred farmers' cultivars, disseminated with pest- and disease-free planting material.

Use rates are very low overall for the new hybrids diffused in Uganda. Previous field research suggests that the cooking quality is considered inferior for hybrids, which is also reinforced by our findings (shown in Table 5.11). Two aspects of diffusion may also provide some explanation for low overall use rates. First, improved bananas were disseminated by publicly funded programs in both Uganda and Tanzania, but appear to have been targeted in Uganda to fewer communities, most located in the lowlands. Consequently, only one-third of the villages surveyed in Uganda (9 of 27) had any adoption, while within those communities, up to 65 percent of farmers grew exotic bananas or hybrids (see Appendix C). A second aspect of diffusion is that there is significant use of introduced bananas in nonexposed areas, where no program or project introductions had been made. The importance of farmer-to-farmer transfers of planting material for bananas, a vegetatively propagated crop, is salient.

In Tanzania, hybrid adoption rates are substantially higher than in Uganda, attaining about 19 percent of farmers in Kagera Region during the survey year. Use of exotic materials is also much greater (Tables 5.15 and 5.16; for names, see Appendix B). Again, however, the proportion of farmers growing hybrids was comparable between areas that were exposed and those that were not exposed,[5] attesting to

[5] The terms are defined in Chapter 2. "Exposure" means that the area has received introductions of planting material through a formal program or has been directly influenced by a program activity.

Table 5.14 Number and share of banana mats planted to improved banana cultivars, by users, Uganda

	Elevation		Exposure		
Cultivar type	Low	High	Exposed	Not exposed	All
Number of mats					
Exotic or hybrid	7.9	0.0	5.4***	8.3***	7.9
Elite endemic	9.2	0.0	9.2***	6.7***	9.2
All improved	17.0**	0.0**	17.0***	15.0***	17.04
Mat share					
Exotic or hybrid	0.03**	0.0**	0.1***	0.02***	0.03
Elite endemic	0.1*	0.0*	0.2**	0.1**	0.10
All improved	0.13**	0.0**	0.3***	0.12***	0.13

Notes: ***, **, and * indicate statistical significance at the 1 percent, 5 percent, and 10 percent levels, respectively, in the difference of means or distributions across elevation and exposure strata within country. Mean number of mats in improved material is calculated only for those growing the improved material. A mat is a mother plant and all of its plantlets. Exotic refers to introduced farmers' cultivars; elite endemic refers to endemic types selected and enhanced by the National Agricultural Research Organization; and all improved combines exotic, elite endemic, and hybrid categories.

Table 5.15 Percentage of farmers growing improved banana cultivars, Tanzania

	Elevation		Exposure		
Type of cultivar	Low	High	Exposed	Not exposed	All
Exotic	97.3	100	97.3	100	97.7
Hybrid	21.4***	5.0***	18.6	20	18.9
Either exotic or hybrid	97.3	100	97.3	100	97.7

Notes: Differences across strata are tested by cross-tabulation using the chi-square distribution. *** indicates statistical significance at the 1 percent level. Columns do not total because 35 households were found growing both exotics and hybrids on their plots, reducing the total number of households growing at least one improved cultivar to 132 instead of 174. Exotic refers to introduced farmers' cultivars.

the effects of farmer-to-farmer diffusion. This finding is consistent with the pattern estimated by the KCDP project, in Table 4.3 of Chapter 4. Farmer-based exchange more than offset project-based exchange, even when project efforts were as intensive as those undertaken in Kagera Region.

Use of recommended banana-management practices was studied only in Uganda (Katungi 2006). Use rates for recommended management practices are much higher than for improved cultivars (Table 5.17). However, farmers use a wide range of organic fertilizers and sanitation practices to manage their banana groves. They apply mulch (in the form of grass, crop residues, or household refuse) to control weeds, enable water penetration, and add nutrients to the soil when decomposed. Other organic fertilizers are applied directly to the soil as organic manure (that is, animal waste and composted household refuse). On average, mulching, animal manure, and soil and water conservation (SWC) bands are more

Table 5.16 Number and share of banana mats planted to improved banana cultivars, by users, Tanzania

	Elevation		Exposure		
Cultivar type	Low	High	Exposed	Not exposed	All
Number of mats					
Exotic	16.4***	51.9***	17.4**	40.8**	25.2
Hybrid	5.6***	22.5***	6.4	4.9	6.0
Exotic or hybrid	16.4***	51.9***	17.4**	40.8**	25.2
Mat share					
Exotic	0.40***	0.14***	0.30**	0.38**	0.33
Hybrid	0.01**	0.001**	0.01	0.01	0.01
Exotic or hybrid	0.41***	0.14***	0.31**	0.39**	0.34

Notes: *** and ** indicate statistical significance at the 1 percent and 5 percent levels, respectively, in the difference of means or distributions across elevation and exposure strata within country. Exotic refers to introduced farmers' cultivars.

Table 5.17 Percentage of farmers adopting selected banana management practices, Uganda

	Elevation		Exposure		
Practice	Low	High	Exposed	Not exposed	All
Natural resource management					
Mulching	66.5***	91.9***	73.5**	59.4**	75.1
Animal manure	35.6***	64.6***	37.8	33.3	45.4
Composting	17.01	15.2	26.5***	7.3***	16.4
Soil and water conservation contour bands	9.8	15.2	16.3***	3.1***	11.6
Soil and water conservation other bands	12.4***	30.3***	19.4***	5.2***	18.4
Mat management (sanitation practices)					
Corm paring	18.9	14.3	33.0***	6.1***	17.3
Desuckering	73.5***	92.9***	79.4*	67.7*	79.9
Detrashing	95.4**	98.9**	97.9*	92.9*	95.6
Stumping	64.3***	98.9***	68.0	60.6	75.9
Splitting/chopping pseudostem	65.3***	79.6***	67.0	63.6	75.9
Corm removal	50	80.6	57.0***	34.3***	60.2
Weevil trapping	31.6	24.5	37.1	26.3	29.3

Note: ***, **, and * indicate statistical significance at the 1 percent, 5 percent, and 10 percent levels, respectively, in the difference of means or distributions across elevation and exposure strata within country.

Table 5.18 Share of banana mats managed with recommended practices among users, Uganda

	Elevation		Exposure		
Practice	Low	High	Exposed	Not exposed	All
Natural resource management					
Mats under mulch	0.35***	0.12***	0.38*	0.29*	0.26
Mats under animal manure	0.30***	0.12***	0.33	0.24	0.22
Mats under compost	0.30***	0.11***	0.31	0.26	0.25
Mat management					
Proportion of stumps removed	0.47***	0.64***	0.56***	0.35***	0.54
Proportion of pseudostems split or chopped	0.26***	0.59***	0.33***	0.16***	0.38

Note: *** and * indicate statistical significance at the 1 percent and 10 percent levels, respectively, in the difference of means or distributions across elevation and exposure strata within countries.

frequently used in high-elevation areas (which are also in the southwest) than in low-elevation areas (Central and Eastern regions). Use rates of most sanitation practices (that is, desuckering, detrashing, stumping, and splitting/chopping pseudostem) were also significantly higher in high-elevation areas than in lowlands. In general, there is a significant difference in the management of bananas between the two major producing areas defined by altitude, probably due to differences in the production potential. Farmers in high-altitude areas have a greater incentive to manage their plants well, partly due to the more commercial nature of production. Rates of mulching, manure application, and water conservation reflect resource availability and, for bands, whether the planting is on slopes.

Statistical differences in use of specific management practices are also evident according to exposure, although not for all practices. Adoption rates for mulching, composting, SWC bands, corm paring, detrashing, desuckering, and corm removal were higher in exposed areas than in nonexposed areas. The adoption rates of SWC contour bands and stumping in low-altitude areas were observed to be slightly higher among relatively small holders (<2 ha), with no significant differences based on size for other practices. Higher adoption rates for most of the banana management practices were found near paved roads, suggesting that improved access to markets and information may increase the diffusion of banana management practices.

Although many farmers report use of soil fertility management practices, the extent of use per grove among the adopters is relatively low (Table 5.18). Significant differences in the number of mats grown under mulch (grass or crop residues) between the two elevation areas were found. The share of the plantation under all three organic fertilizers was higher in low-altitude areas than in highlands, implying a more binding constraint regarding access to the organic fertilizers for farmers in high-altitude areas, who are also comparatively large-scale banana farmers. There are no meaningful differences in natural resource management practices by exposure, which suggests that elevation is a more important underlying parameter, because it is related to soil fertility, pest and disease pressures, and the opportunity cost of labor (Bagamba, 2007).

The extent of use of sanitation practices (measured by the extent of stumping and pseudostem splitting or chopping) was also significantly higher in high-elevation areas compared to low-elevation areas, and in exposed areas relative to nonexposed areas.

Table 5.19 Average number of years in use of banana cultivars currently planted, Uganda

	Elevation		Exposure		
Type of cultivar	Low	High	Exposed	Not exposed	All
Established in village	10.9***	26.3***	17.5***	12.5***	13.2
Introduced to village	7.8*	13.5*	4.7**	8.8**	8.1
All cultivars grown	10.6***	25.5***	16.5***	12.1***	12.8

Notes: Total number of households in all strata is 514. Total number of cultivars in all strata is 3,713. ***, **, and * indicate statistical significance at the 1 percent, 5 percent, and 10 percent levels, respectively, in the difference of means across elevation and exposure strata within countries.

About 64 percent of the lower pseudostems[6] are chopped off a few days after harvest, compared to 47 percent in low-elevation areas. Similarly, 59 percent of the upper pseudostems in high-elevation areas are either split or chopped soon after harvest, whereas only 26 percent of the upper pseudostems in the lowlands are managed accordingly.

Banana Planting Material Systems

The average numbers of years in use for both endemic cultivars that farmers consider to be "ancestral" (long-established in their village) and cultivars that have been introduced are much greater in the highlands than in lower elevations (Table 5.19). In the highlands, introduced cultivars include only farmers' cultivars. In fact, banana plants last longer in the highlands. In the lowlands, farmers shift and replant bananas more frequently but over their lifetimes, they have a similar cumulative experience growing the crop. Villages where at least some farmers have been exposed to new cultivars are mostly found in the lowlands. There, farmers have grown introduced cultivars, whether improved or not, for a shorter average period of time. This pattern likely reflects the greater biotic pressures and farmers' recognition of the need to reintroduce disease- and pest-free planting material. Among all farmers surveyed, the average number of years in use of a banana cultivar is about 13 in general, and is 8 for those introduced to the village, independent of improvement status.

In Tanzania, the average number of years a cultivar has been grown by farmers ranged from 1 to 54 years. Few farmers have maintained the same cultivar for more than 30 years. At the same time, some farmers appear to be replanting older, ancestral cultivars. Data on historical cultivation of banana cultivars reveal that over the past 15 years, on average, households have abandoned four banana cultivars in the Bukoban and Karagwe Ankolean lowlands and two banana cultivars in the Karagwe Ankolean highlands. The genomic group with the highest number of abandoned cultivars is the endemic group, but at the lowest rate. All genomic groups, ancestral and introduced cultivars, are represented among those discarded by farmers. Farmers reported a range of reasons for their decisions, including those related to preferences, markets, and climate.

[6] Bananas are normally harvested about 1 m above the ground; the part that remains is referred to here as "lower pseudostem," and the harvested part is referred to as "upper pseudostem." After the fruit is cut off, the lower part of the pseudostem is also cut off, which is called "stumping," while the upper part of the pseudostem can either be peeled or chopped to destroy breeding grounds for pests and also facilitate quick decomposition to recycle nutrients taken up by the plant during its growth.

Table 5.20 Average number of years in use of banana cultivars currently planted, Tanzania

	Elevation		Exposure		
Type of cultivar	Low	High	Exposed	Not exposed	All
Endemic	20.1	23.5	20.6	21.6	21.0
Nonendemic					
Exotic	18.7***	23.3	19.5	20.7	19.9
Hybrid	2.5	2.0	2.2	3.5	2.5
All cultivars grown	20.7	23.9	21.0	22.3	21.5

Note: *** indicates statistical significance at the 1 percent percent level in the difference of means or distributions across elevations within countries.

Endemic cultivars have been grown the longest of all genomic groups (20 years), compared to 3 years for hybrids recently introduced (Table 5.20). Cultivar ages are generally older in the high elevations, where disease pressures are less and farmers do not need to replace materials as frequently to maintain yields. Exotic and hybrid cultivars are the primary options in areas where endemic bananas are no longer performing well. Older cultivar age is even more pronounced for the Karagwe highlands, where no exotic cultivars were recorded and banana hybrids were introduced most recently.

Table 5.21 provides additional evidence that transfer of planting material is much more intensive in low- than in high-elevation areas of Uganda, in terms of several indicators: (1) the larger percentage of farmers supplying and receiving banana planting material, (2) the large volume of planting material supplied and received, and (3) the high average replacement ratios. Similar patterns appear for farmers in exposed areas, found primarily in low-elevation areas. Supplied material is defined as all banana planting material that farmers have supplied from the different cultivars they grow. Received material is defined as banana suckers of only those cultivars that the farmer considers introduced to the village, rather than ancestral (that is, established) in the village. The replacement ratio is defined as the number of suckers received per number of years growing the cultivar.

The most frequently supplied cultivars in farmer-to-farmer transactions are endemic cooking cultivars. When the quantity of suckers supplied is taken into account, the planting material of nonendemic and noncooking banana cultivars (including banana hybrids) dominates transactions. These are sold for cash. The most frequently received cultivars are also endemic cooking types. Banana hybrids dominate the planting materials received by farmers for cash by a large margin.

On average, farmers in low-elevation areas traverse a greater distance to get to the point of supply, and similarly for farmers in exposed areas (Table 5.22). For farmers in nonexposed areas, the mean distance is much larger than for those in exposed areas. On average it takes 15 km for farmers interested in acquiring planting material to reach the point of acquisition, evidence of large transaction cost given the limited means of transportation of these farmers.

The vast majority of planting material changes hands without the exchange of money (Table 5.23). A greater proportion of planting material is supplied for cash in low- than in high-elevation areas, and in exposed compared to nonexposed areas.

That planting material is typically exchanged without the use of money was also

Table 5.21 Planting material transfer (supplied and received) for cultivars grown in the survey season, Uganda

	Elevation		Exposure		
	Low	High	Exposed	Not exposed	All
	Percentage of banana farmers				
Supplying materials within or outside village	66.8***	32.7***	57.8	58.7	60.3
Receiving materials introduced to the village	20.2***	6.2***	23.7	14.5	17.6
	Quantity of planting material				
Total number of suckers supplied within or outside village	22,217	602	12,067	10,752	22,819
Total number of suckers received that were introduced to the village	2,250	412	1,780	882	2,662
Replacement ratio	6.1***	0.9***	19.1***	3.0***	5.8

Note: *** indicates statistical significance at the 1 percent level in the difference of means and distributions across elevation and exposure strata within countries.

Table 5.22 Planting material supplied or received, by distance, social relationship, elevation, and exposure, Uganda

	Elevation		Exposure		
	Low	High	Exposed	Not exposed	All
Supplied material					
Mean distance to point of supply (km)	1.2***	0.9***	2.1***	0.9***	1.1
Most frequent social relationship to recipient (occupation)	Friend (farmer)	Friend (farmer)	Friend (farmer)	Friend (farmer)	Friend (farmer)
Received material					
Mean distance to point of acquisition (km)	15.3	12.7	4.7***	17.5***	15.2
Most frequent social relationship to source (occupation)	Friend (farmer)	Friend (farmer)	Friend (farmer)	Friend (farmer)	Friend (farmer)

Notes: For planting material supplied, the information is computed over all cultivars listed as having been transferred ($N = 1{,}253$). For planting material received, the information is computed over all cultivars listed as having been introduced to the village ($N = 265$). *** indicates statistical significance at the 1 percent level in the difference of means across elevation and exposure strata.

Table 5.23 Percentage of planting material supplied and received, by type of transaction and stratum, Uganda

	Elevation		Exposure		
Transaction type	Low	High	Exposed	Not exposed	All
Supplied material					
For cash	10.6	21.4	22.7	5.6	11.3
Not for cash	89.1	78.7	77.3	94.0	88.4
Received material					
Cash payment	24.0	0.0	27.5	14.3	23.0
No cash payment	76.0	100.0	72.5	85.7	77.0

Notes: Computed over all cultivars listed as having had transfers or receipts. The total number of observations for supplied planting material is $N = 1{,}29\ 3$, while the total number of observations for received planting material is $N = 244$. Some of the percentages do not add up because of missing values.

Table 5.24 Average farm-gate and market bunch prices across banana cultivars (Ush)

	Elevation		Exposure		
Bunch price	Low	High	Exposed	Not exposed	All
Farm gate	1,296***	1,999***	1,203***	1,516***	1,425
Market	2,328	4,363	1,709	2,689	2,649

Notes: The total number of observations across cultivars in all strata for farm-gate price is $N = 914$, and for market price it is $N = 187$. *** indicates statistical significance at the 1 percent level in the difference of means across elevation and exposure strata.

made evident in farmers' responses to hypothetical questions about the amounts they would be willing to pay or accept in cash for planting materials. Little variation in their responses was observable, and where variation was observed, it was associated with respondents rather than cultivars.

Differences in the values of cultivars are better reflected in output prices for banana bunches at the farm gate or in the market. Banana bunches differ in size, cluster composition, size and shape of fruit, and visible indicators of consumption quality, such as cooking quality. The average price of banana bunches at the farm gate and market is reported in Table 5.24.

The data also indicate that market margins are high, particularly in high-elevation areas, beause of the distance to main urban markets and related transportation and handling costs. The average farm-gate price is higher in the highlands than at low elevations. This result can be explained by the commercially oriented production of the highlands, where biotic pressures are lower and farmers manage their plantations more carefully. Banana bunches originating from the high-elevation areas of southwest Uganda are generally larger and are reported to taste better.

Other farmers are also the dominant source of banana planting material in Tanzania (Table 5.25). Extension agents and NGOs were next in importance. Selling of

Table 5.25 Number of banana suckers received, by source and genomic group, Tanzania

Genomic group	Farmer	Trader	Extension agent	Research	Nongovernmental organizations	Community garden
Endemic	1,232	7	0	2	5	17
Nonendemic	513	9	1	0	6	3
Exotic	130	4	32	1	22	0
Hybrid	24	1	29	2	24	0
Total	1,899	21	62	5	57	20

banana suckers by either traders or community gardens was relatively less important as a source of planting material.

The estimated distance from the source of planting materials (banana suckers) ranged from 0.01 to 88 km, with an average distance for all types of 3 km (Table 5.26). The low average relative to the maximum indicates that most farmers obtained planting material within or near their villages. Planting material of cultivars not widely cultivated were obtained at greater distances.

The planting materials of exotic cultivars were found at the greatest distances, followed by hybrid, nonendemic, and endemic cultivars. This order reflects the availability of planting materials by genomic group. Farmers located at high elevations obtained their endemic and exotic planting materials at shorter distances than those located at low elevations, but the former had to travel farther for hybrids. Ensuring pest- and disease-free planting materials is a major motivation of farmer behavior. Planting materials in the nonexposed areas were found at greater distances than in the exposed areas.

In Tanzania, as in Uganda, banana planting material exchange among farmers generally does not involve monetary transactions (Table 5.27). Five means of acquiring planting material were identified: gift, aid, purchase, exchange, and free of charge from extension sevices. Seventy-eight percent of farmers acquired the planting material of improved banana cultivars free of charge. About 11 percent of farmers acquired their planting material by purchase. More than 95 percent of endemic cultivars were acquired free of charge, only 3 percent by purchase, and the remainder by aid or gift.

Social relationship plays a key role in the exchange of planting material (Table 5.28). About 35 percent of the farmers obtained their planting material from farmers with whom they had no relationship, and friends or family accounted for another 35 percent of material transfers. Work relationship accounted for the remaining 30 percent. Most of those who received materials from someone with whom they had no relationship are found in the lowlands in exposed areas. These materials are likely to be the materials introduced formally.

The average quantities per genomic group of planting materials obtained by farmers reflected their proportion in the total banana mats (Table 5.29). The highest proportion is endemic bananas, followed by exotic cultivars and hybrids. Per household, the average total number of suckers obtained from KCDP (including both hybrids and exotic cultivars) was 2.5. Farmers at high elevations obtained more suckers of endemic bananas and fewer of exotic bananas than those located in the lowlands, where disease pressures are greater. There is clearly a preference for larger stocks of nonendemic planting material among households in areas with binding biotic constraints. Though no significant differences in the demand for hybrid planting material are evident, the distribution of the demand

Table 5.26 Average distance from the source of planting materials, by genomic group and strata, Tanzania (km)

	Elevation		Exposure		
Genomic group	Low	High	Exposed	Not exposed	All
Endemic	2.2	1.1	1.6	1.7	1.9
Nonendemic					
Exotic	6.2	0.5	5.8	8.4	6.1
Hybrid	4.2	5.0	3.7	7.8	4.2
All genomic groups	3.2	1.3	2.5	8.5	2.9

Table 5.27 Percentage of improved banana planting materials obtained, by mode of transaction, Tanzania

	Elevation		Exposure		
Means of acquisition	Low	High	Exposed	Not exposed	All
Gift	5.0	0.0	6.0	0.0	4.9
Aid	5.6	0.0	6.8	0.0	5.5
Purchase	11.2	0.0	6.8	30.0	11.0
Exchange	0.6	0.0	0.8	0.0	0.6
Free of charge	77.6	100.0	79.7	70.0	77.9
Total	100.0	100.0	100.0	100.0	100.0

Table 5.28 Percentage of improved banana planting materials received, by social relationship to source, Tanzania

	Elevation		Exposure		
Social relationship	Low	High	Exposed	Not exposed	All
Family	2.5	0.0	2.3	3.3	2.5
Friend	33.1	50.0	26.4	63.3	33.3
Work	29.3	50.0	30.2	26.7	29.6
None	35.0	0.0	41.1	6.7	34.6
Total	100.0	100.0	100.0	100.0	100.0

for other nonendemic material indicates longer use of hybrids in affected areas.

The amount of banana planting material supplied was far greater in the highlands than the lowlands, and in the nonexposed as compared to the exposed areas (Table 5.30). Households in highlands supplied five times the amount of planting material. Because the incidence and severity of biotic constraints in these areas is low, the stock of established planting material is larger and more likely to be free of

Table 5.29 Average number of banana suckers received by farmers per genomic group, by elevation and exposure, Tanzania

	Elevation		Exposure		
Genomic group	Low	High	Exposed	Not exposed	All
Endemic	27.2*	31.2*	40.3**	3.2**	28.2
Nonendemic	17.4**	11.1**	21.0***	5.5***	15.9
Exotic	16.3**	10.8**	20.0***	4.9***	15.0
Hybrid	1.1	0.3	1.0	0.6	0.9
Overall average	44.6**	42.3***	61.3***	8.6***	44.0

Note: ***, **, and * indicate statistical significance at the 1 percent, 5 percent, and 10 percent levels, respectively, in the difference of means or distributions across elevations within countries.

Table 5.30 Average number of banana suckers per genomic group supplied by farmers to other farmers, by elevation and exposure, Tanzania

	Elevation		Exposure		
Genomic group	Low	High	Exposed	Not exposed	All
Endemic	61.8***	373.1***	69.7***	274.9***	137.0
Nonendemic	17.8***	66.5***	15.1***	59.3***	29.6
Exotic	17.4***	66.2***	14.8***	58.6***	29.2
Hybrid	0.4	0.3	0.3	0.6	0.4
All groups	79.7***	439.5***	84.8***	335.0***	166.6

Note: *** indicates statistical significance at the 1 percent level in the difference of means or distributions across elevation and exposure strata within countries.

disease, or at least perceived to be so. The number of hybrids grown in these areas is negligible, both because of less exposure to them and less disease pressure. The exchange of planting material in Tanzania between the lowlands and the highlands, which is driven by the incidence and severity of biotic pressures and other production constraints (such as soil fertility), has implications for program interventions with improved planting material.

Conclusion

Few meaningful differences in household characteristics emerge among banana-growing households according to elevation or location. In Uganda, roughly a quarter of sampled households are self-sufficient in banana production, with most of the remaining households in high-elevation areas participating in banana markets as sellers. Although the proportion of self-sufficient households is higher in Tanzania, most of the households participating in markets sell bananas, particularly in high-elevation areas, a finding that parallels the evidence from Uganda. In Uganda, women's involvement in banana sales is greater in high-elevation areas, with only partial involvement in choosing cultivars or making decisions about production management. Still, men and women participate at roughly equal rates in sales of cooking, roasting, and dessert bananas. Participation of women in banana production, management, and sale is considerably higher in lowland areas of Tanzania, and differs markedly in pattern between banana-growing households in Uganda and Tanzania.

The diversity of banana cultivars is high at the country, village, and household levels across the Lake Victoria region of Uganda and Tanzania. Even the most popular banana cultivars occupy less than 10 percent of the total number of mats grown in Uganda. Cooking bananas are not only the most important food staple for households, but they are also a source of cash in local markets. In Uganda, the findings support banana researchers' observations that (1) farmers have difficulties recognizing and attributing different disease pressures and (2) farmers consider endemic cooking bananas to be superior in terms of cooking quality compared to hybrid cultivars. Furthermore, use rates for recommended management practices appear to be much higher than for improved banana cultivars, particularly in high-elevation and exposed areas. Transfer of planting material, predominantly between farmers, is evidently more intensive in low-elevation areas, where disease pressures are greater. Village social structures and participation in social organizations emerge as important determinants of use of banana technologies.

The number of distinct banana cultivars identified in Tanzania is larger than in Uganda, considering the smaller geographic extent of the Tanzanian sample, and recent introductions of exotic and hybrid bananas have no doubt contributed to the richness of materials grown. A few common characteristics emerge between findings in Uganda and Tanzania, however: (1) the vast majority of the identified cultivars (more than 80 percent) are endemic to the region; (2) on average, households grow numerous distinct cultivars simultaneously in their banana groves; (3) all farms grow at least one cooking cultivar; (4) farmers perceive endemic bananas to be superior in terms of consumption quality; and (5) there is strong evidence of farmer-to-farmer exchange of planting material of both endemic and exotic and/or hybrid cultivars.

Adoption of hybrids and exotic cultivars (both endemic to other regions but not this one, and nonendemic) is greater in Tanzania than in Uganda. The relative severity of biotic pressures in Kagera Region is one reason. Data also confirm that farmers in Kagera Region perceive that hybrids are superior to endemic types in terms of yield and resistance to black Sigatoka and weevils, but inferior with respect to *Fusarium* wilt, as the scientists who recommended the varieties had expected. The same cannot be said for the farmers surveyed in Uganda. Although this result could be an artifact of the relatively small sample of hybrid adopters in Uganda, it is consistent with national researchers' observations that disease pressures were greater in Tanzania, and Ugandan farmers have difficulty in disease recognition. Clearly, dissemination pathways, both formal and informal, are another major factor that explains differences in adoption between the two countries. The next three chapters investigate the determinants and effects of adoption of cultivars and practices.

References

Bagamba, F. 2007. Market access and agricultural production: The case of banana production in Uganda. Ph.D. thesis, Wageningen University, The Netherlands.

Edmeades, S., M. Smale, M. Renkow, and D. Phaneuf. 2004. *Cultivar demand within the framework of an agricultural household model with attributes: The case of bananas in Uganda.* Environment and Production Technology Division Discussion Paper 125. Washington, D.C.: International Food Policy Research Institute.

Gold, C. S., A. Kiggundu, D. Karamura, and A. Abera. 1998. Diversity, distribution and selection criteria of *Musa* germplasm in Uganda. In *Bananas and food security,* ed. C. Picq, E. Fouré, and E. A. Frison. International symposium on the socioeconomics of non-export banana production, Douala, Cameroon, November 10–14, 1988. Montpellier, France: International Network for the Improvement of Banana and Plantains.

Gold, C. S., G. Kagezi, G. Night, and P. Ragama. 2004. The effects of banana weevil damage on highland banana growth, yield and stand duration in Uganda. *Annals of Applied Biology* 145 (3): 263–269.

Hardaker, J., R. Huirne, and J. Anderson. 1997. *Coping with risk in agriculture.* Wallingford, U.K.: CAB International.

Katungi, E. 2006. Social capital and technology adoption on small farms: A case of banana production technology in Uganda. Ph.D. thesis, University of Pretoria, Pretoria, Republic of South Africa.

Nkonya, E., J. Pender, C. Kaizzi, K. Edward, and S. Mugarura. 2005. *Policy options for increasing crop productivity and reducing soil nutrient depletion and poverty in Uganda.* Environment and Production Technology Division Discussion Paper 134. Washington, D.C.: International Food Policy Research Institute.

Nkuba, J. 2007. Assessing adoption and economic impacts of new banana cultivars on livelihoods of farmers in Kagera Region, Tanzania. Ph.D. thesis, Department of Agricultural Economics, Sokoine University, Sokoine, Tanzania.

Part III. Economic Assessment of Technology Adoption and Impact

CHAPTER 6

A Trait-Based Model of the Potential Demand for Transgenic Banana Cultivars in Uganda

Svetlana Edmeades and Melinda Smale

This chapter derives farmer demand for planting material from the demand for genetic traits and cultivar attributes in an agricultural household model with missing markets, building on the general model presented in Chapter 2.[1] Applied to the sample data described in Chapter 2 and Appendix D, the econometric approach is used to explore several issues of importance to national decisionmakers. We begin by identifying the determinants of farmer demand for planting material of cooking banana cultivars that are candidate hosts for insertion of pest and disease resistance by NARO. At the level of the individual farm household, demand for planting material is defined by the decision to "adopt" (use) and the scale of use (number of mats). We then use the fitted equation to generate three pieces of information. First, we develop client prototypes by describing the characteristics of farm households with high and low predicted demands for particular cultivars. This exercise illustrates how the choice of host planting material for genetic transformation can have social consequences. Second, we use the example of Nakitembe to estimate the total size of industry demand for planting material of a genetically transformed host cultivar. Third, we simulate changes in demand for planting material when resistance traits are inserted, with varying degrees of effectiveness, and when other supporting public investments are made in extension, market infrastructure, and education. The simulation demonstrates how the magnitude of the payoff to research investment crucially depends on other types of investments. This point reiterates that, as documented in other literature about agricultural innovations (see Chapter 1), seed-based technical change depends not only on the release of an improved cultivar, but also on the markets, information, and policies that enable its widespread adoption.

Theoretical Model

The trait-based model in this chapter (Edmeades 2003) derives the demand for planting material in the decisionmaking framework of the agricultural household with imperfect markets (Singh, Squire, and Strauss 1986). The conceptual framework draws from Lancaster's theory of consumer choice (1966) and models of demand for farm input and output characteristics (Ladd and Martin 1976; Ladd and Suvannunt 1976). The model focuses on the demand for planting material, or seed in the broad sense, including endemic farmers' cultivars, cultivars improved through crossing, and genetically engineered cultivars as special cases.

---

[1] A more detailed version of this chapter was published in *Agricultural Economics* in 2006.

The agricultural household maximizes utility (u) from the set of intrinsic attributes ($\mathbf{z}^C$) of the goods it consumes ($\mathbf{x}$), the consumption of other goods (x^G), and leisure or home time (h), given a set of characteristics describing the composition of the farm household ($\mathbf{\Omega}_{HH}$) and the local market conditions ($\mathbf{\Omega}_M$). The household chooses the type and amount of consumption goods ($\mathbf{x}$), the type and amount of planting material ($\mathbf{v}$), other goods (x^G), and labor hours (l) dedicated to production of the crop:

$$\max_{\mathbf{x}, x^G, \mathbf{v}, l} u\left[\mathbf{x}(\mathbf{z}^C), x^G, h \mid \mathbf{\Omega}_{HH}, \mathbf{\Omega}_M\right]. \qquad (1)$$

The household maximizes utility subject to a number of constraints. The production technology is defined by variable inputs, including $\mathbf{v}$ and l, used for the production from each cultivar ($\mathbf{q}$) in a banana grove of fixed size in the short term. Each cultivar has an expected level of agronomic traits ($\mathbf{z}^P$), as perceived by the farmer. Although the bundles and levels of attributes provided by cultivars are fixed from the perspective of an individual household, the household can vary the type and amount of consumption and production attributes by changing the combination of cultivars and quantities of planting material grown:

$$g\left[\mathbf{q}, \mathbf{v}(\mathbf{z}^P), l \mid \mathbf{\Omega}_F, \mathbf{\Omega}_M\right] \leq 0. \qquad (2)$$

The production technology is conditioned on the physical characteristics of the farm, denoted by $\mathbf{\Omega}_F$. The vector of market characteristics ($\mathbf{\Omega}_M$) is included in both the utility and production functions because market imperfections potentially affect both demand and supply in the agricultural household.

Household budget limitations are depicted by the full income constraint, where $\mathbf{p}$ is a vector of crop product prices, p^G is the price of other goods consumed by the household, and I is exogenous income:

$$(\mathbf{q} - \mathbf{x})' \mathbf{p} - p^G x^G + I \geq 0. \qquad (3)$$

The full income constraint is defined over all tradable crop products, meaning that a market exists for these products, and households choose to transact in it.

The participation of household members in market transactions depends on market existence and completeness, and the type and magnitude of transactions costs they encounter (de Janvry, Fafchamps, and Sadoulet 1991). Although output markets for crop products exist in rural Uganda, they are often incomplete, failing to capture quality differentials among cultivars. Moreover, households are often located far from output markets. Product market failure underscores the importance of meeting immediate household consumption demand through own production. Specific to households, this situation is reflected in the nontradability constraint:

$$\mathbf{q} - \mathbf{x} = 0. \qquad (4)$$

In addition, input markets are often missing. For many households, production decisions are likely to be influenced by endogenous values rather than by exogenous market prices of inputs. The primary source of labor for crop production is typically the family. The time constraint expresses the distribution of total available time (T) between production and home activities:

$$T - l - h = 0. \qquad (5)$$

Planting material is either reproduced on-farm or obtained through farmer-to-farmer exchange. The planting material constraint states that the cultivar choices of farmers are limited by the range of traits and attributes available to them locally. The number of distinct cultivars existing in the village, denoted by $\tilde{v}$, represents the local stock of cultivar attributes. Households grow only those cultivars available to them. For those cultivars not in the feasible set, the revealed demand for planting material of a cultivar v_i is 0:

$$\sum_{i \notin \tilde{v}} v_i = 0. \tag{6}$$

Corner (zero) solutions are possible for specific cultivars, because the set of planted cultivars need not be the same across households. The scale constraint circumscribes the size of the banana grove, where $\bar{v}$ denotes the total number of mats, measured in the same units as:

$$\bar{v} \geq \sum_{i \in \tilde{v}} v_i. \tag{7}$$

A set of non-negativity restrictions are imposed:

$$x_i \geq 0, q_i \geq 0, v_i \geq 0. \tag{8}$$

Kuhn-Tucker conditions are used to solve for v_i, the demand for planting material from cultivar $i = 1, \ldots, N$:

$$v_i = v_i(\mathbf{z}^C, \mathbf{z}^P, \mathbf{p}, p^G, I, T, \bar{v}, \tilde{v} \mid \boldsymbol{\Omega}_{HH}, \boldsymbol{\Omega}_F, \boldsymbol{\Omega}_M), \tag{9}$$

where $v_i = 0\ \forall i \notin \tilde{v}$, and $v_i = v_i(\cdot)\ \forall i \in \tilde{v}$ such that $\omega \subset \tilde{v} : v_i > 0\ \forall i \in \omega$ and $s \subset \tilde{v} : v_i = 0\ \forall i \in s$, where ω denotes the subset of cultivars from the feasible set $\tilde{v}$ that the household chooses to grow in positive numbers, and s depicts a subset of cultivars available but not grown by the household.

The demand for planting material of a cultivar (v_i) is defined as the number of banana mats of cultivar i grown by the household. Bananas are grown in a mat consisting of the mother plant and plantlets and are vegetatively propagated from a plantlet. Thus, the demand for a cultivar is expressed as a count of plants rather than as a share of crop area, as has often been the case in the literature about adoption of improved cereal cultivars. The demand for a cultivar is determined by the bundle and levels of consumption attributes and agronomic traits it provides ($\mathbf{z}^C$,$\mathbf{z}^P$), household characteristics ($\boldsymbol{\Omega}_{HH}$, T,I), farm characteristics ($\boldsymbol{\Omega}_F$,$\bar{v}$,$\tilde{v}$), and market characteristics ($\mathbf{p}^B$, p^G, $\boldsymbol{\Omega}_M$), where $\mathbf{p}^B$ are banana bunch prices.

Observed outcomes are a combination of interior and corner solutions. Corner solutions arise for two reasons. One occurs when cultivars are not locally available to households, and it is exogenous ($\forall i \notin \tilde{v}$). A second reason is endogenous, occurring when cultivars are available in the village, but households choose not to grow them ($\forall i \in s$). Each household chooses to grow a subset of available cultivars ($\forall i \in \omega$) in positive numbers and at different levels. The solution to the optimization problem consists of two components: the choice to grow a subset of available cultivars and their scale of cultivation, or mat count. Because households often grow several cultivars simultaneously to meet their demand for attributes and traits, the empirical analysis is based on a system of censored equations, as defined in equation (9).

Econometric Approach

A limitation of most adoption studies is that the censored dependent variable is defined over actual choices only for cultivars that are currently grown, creating a potential for selection bias. This limitation is generally imposed by the underlying data. The data collected in Uganda allow for the recovery of a more complete choice set of banana cultivars by combining both revealed and stated preferences for the attributes of cultivars that are currently grown and no longer grown, but known to the farmer.

The count of banana mats is zero for cultivars not in the feasible set ($\forall i \notin \tilde{v}$). Cultivars not in the feasible set are those to which farmers have never been exposed: they have neither observed nor grown such cultivars. The lack of exposure to these cultivars precludes household awareness of their attributes. Farmers are aware of the attributes of the cultivars they currently grow in positive numbers ($\omega \subset \tilde{v} : v_i > 0$). Farmers are also familiar with the attributes of cultivars they do not currently grow, but have grown in the past or currently observe in the banana groves of their neighbors ($s \subset \tilde{v}: v_i = 0$).

The definition of the dependent variable imposes a unique shape on the underlying distribution, which is strongly skewed to the right. The concentration of the mass on the corner ("excess zeros") and the formulation of the dependent variable in terms of integer values (count of banana plants) dictate the need for a count data approach (Cameron and Trivedi 1998). One such approach is the zero-inflated Poisson (ZIP) system, which is used to predict the number of mats of cooking banana cultivars grown, defined by the set $\tilde{v}$.[2]

Zero inflated models, often employed to jointly estimate censored systems of equations, characterize the separate mechanisms that generate corner solutions by assigning different probabilities to the observed outcomes based on a logit formulation:

$$F(\gamma_i \mathbf{Z}) = \frac{\exp(\gamma_i \mathbf{Z})}{1 + \exp(\gamma_i \mathbf{Z})}.$$

The vector **Z** depicts the set of exogenous characteristics explaining the probability of each outcome, and γ_i is the vector of parameters to be estimated:

$$P[v_i = 0] = F(\gamma_i \mathbf{Z}) \quad \forall i \notin \tilde{v}$$

$$P[v_i \sim Poisson(\mu_i)] = 1 - F(\gamma_i \mathbf{Z}) \quad \forall i \notin \tilde{v}$$

For all distinct cultivars (and attributes) in the feasible set, the count of banana plants is a non-negative integer distributed Poisson with $\mu_i = \exp(\beta_i \mathbf{X})$. The vector of exogenous characteristics (**X**) includes the consumption attributes and agronomic traits of cultivars (which are not part of **Z**),[3] with other household, farm, and market characteristics. The ZIP formulation accounts for the awareness of cultivar attributes, as well as for levels of cultivar demand.

Demand for planting material is estimated jointly as a system of $i = 1, \ldots, N$ independent censored count equations. Models that treat correlated errors for large systems of censored (count) demand equations have not yet been sufficiently developed for application (Englin, Boxall, and Watson 1998; von Haefen, Phaneuf, and Parsons 2004). In a nonstructural simultaneous system, in any case, accounting for error correlations would serve only to increase estimation efficiency.

Dependent Variables

The spatial diversity of distinct banana cultivars on farms and across the domain is considerable (see Chapter 5 and Appendix A). The dependent variable is defined as the number of mats planted of seven candidate host cultivars ($\tilde{v}$ for $i = 1, \ldots, 7$), representing the revealed demand for planting material and almost half of the total mat numbers of cooking cultivars in the survey domain. A large number of cultivars are observed on household farms in Uganda. No single cooking cultivar occupies more than 9 percent of the total number of banana plants grown by all farmers surveyed. The vast majority of cooking cultivars (75 percent) represent a small share of all cooking banana plants in the sample—each cultivar occupying less than 1 percent. Because farmers grow on average seven different cultivars per plantation, a set of seven cultivars was taken as representative.

The seven cultivars were selected from among 67 potential host cooking cultivars identified in the sample based on a combi-

[2] The ZIP was found to perform better than the simple Poisson model. The Vuong statistic (distributed standard normal) for the test of a ZIP model versus a standard Poisson model is 9.73, which favors the zero-inflated model (Vuong 1989).

[3] The difference in the composition of the sets represented by the vectors X and Z is data driven. Attribute information is available for only the cultivars in the feasible set.

Table 6.1 Summary statistics for variables defined at the household level

Variable	Definition	Mean	Standard deviation
Explanatory variables			
Experience	Ratio of years of experience to age of person in charge of banana production	0.22	0.04
Education	Years of schooling of the person in charge of banana production	5.54	0.72
Dependency ratio	Ratio of household dependents (aged 1–15 and older than 55 years) to total household size	0.47	0.05
Household size	Total number of household members	5.55	0.50
Extension	Number of visits by extension agents in the last 6 months	0.69	0.42
Assets	Value of livestock owned by the household (ten thousand Ush)	43.68	19.40
Exogenous income	Exogenous income the household has received in the previous year (ten thousand Ush)	122.73	60.46
Banana area	Area allocated to banana production (acres)	0.92	0.28
Planting material	Number of distinct banana cultivars available in the village	23.69	1.15
Elevation	Elevation (1 = below 1,400 m.a.s.l.; 0 = above 1,400 m.a.s.l.)	0.89	0.06
Time	Hours to nearest market for bananas	1.07	0.13

Notes: The mean is calculated as a weighted average for each variable. m.a.s.l. is meters above sea level.

nation of scientists' and farmers' criteria. NARO scientists have identified several cultivars for initial transformation assays to represent the range of genomic and use-group diversity found among clone sets in Uganda (Karamura and Pickersgill 1999).[4] Survey data reveal the cultivars that are currently most frequently and extensively grown by farmers in major banana-growing areas of Uganda, suggesting that these are preferred, given the many constraints farmers face and current field conditions. The seven cultivars chosen for the analysis include two that are both widely grown and were initially identified for assays (Mbwazirume and Kibuzi), one that is widely grown but not yet targeted for assays (Nakitembe), and four that are identified, although not as extensively grown (Nakinyika, Enjagata, Kisansa, and Mpologoma). All are endemic cooking cultivars of the East African highlands.

Explanatory variables

The determinants of cultivar demand include household ($\mathbf{\Omega}_{HH}$, I), farm ($\mathbf{\Omega}_F$, $\bar{v}$, $\tilde{v}$) and market characteristics ($\mathbf{p}\ \mathbf{\Omega}_M$) and cultivar attributes ($\mathbf{z}^C$, $\mathbf{z}^P$), as specified in the reduced form equation (9).[5] The comparative statics of a nonseparable agricultural household model are complex, and in general, unambiguous signs on the direction of effects cannot be derived. Hypothesized effects are supported by observations from banana research or the literature on seed innovations in developing economies.

Explanatory variables are defined and summarized at the household level (Table 6.1) and at the cultivar level (Table 6.2). Household characteristics include the expe-

[4] Endemic banana cultivars are classified into five clone sets. The 67 cooking cultivars identified in the sample are variants within four of these clone sets, with the fifth one representing endemic beer cultivars. The seven cooking cultivars used in the analysis span the four clone sets.

[5] No data were available for prices of other goods purchased by the household (p^G). This variable was not included in the analysis.

Table 6.2 Summary statistics for variables defined at the cultivar level

Variable	Definition	Mean (standard deviation)						
		Nakitembe	Mbwazirume	Kibuzi	Nakinyika	Enjagata	Kisansa	Mpologoma
Dependent variable								
Count	Number of plants for $i = 1, \ldots, 7$ banana cultivars	11.60 (6.95)	3.74 (2.29)	3.92 (1.74)	3.16 (1.77)	0.97 (0.68)	2.24 (4.60)	1.70 (4.56)
Explanatory variables								
Farm-gate price	Expected price received at farm gate (thousand Ush), by cultivar	2.05 (0.13)	2.10 (0.15)	1.88 (0.15)	1.69 (0.12)	1.81 (0.12)	2.15 (0.14)	3.03 (0.31)
Cooking quality	Cooking quality (1 = bad; 2 = neither good nor bad; 3 = very good)	2.92 (0.07)	2.91 (0.06)	2.89 (0.07)	2.97 (0.03)	2.93 (0.05)	2.88 (0.06)	2.82 (0.08)
Loss due to black Sigatoka	Expected bunch size loss (percent) (in the case of black Sigatoka)	6.05 (2.01)	4.14 (1.46)	7.90 (2.39)	6.68 (2.49)	1.63 (0.64)	3.53 (1.13)	3.56 (1.61)
Loss due to weevils	Expected bunch size loss (percent) (in the case of weevils)	13.17 (2.87)	13.73 (2.76)	13.90 (2.68)	11.31 (2.56)	8.91 (1.55)	16.17 (3.31)	9.66 (2.27)

Notes: The mean is calculated as a weighted average for each variable. See footnote 9, Chapter 5, for explanation of measurement of yield losses.

rience and education of the banana production decisionmaker,[6] household size, the proportion of dependents in each household, the number of visits by extension agents, livestock assets, and exogenous income (defined as household receipts of remittances and other nonfarm income in the previous year). Larger households and those with more dependents are expected to have greater cultivar demand for meeting immediate consumption needs. No sign is hypothesized a priori for other household characteristics.

Farm characteristics are banana production area, household location (by elevation), and the number of distinct cultivars available in the village. Given the taxonomy used to classify cultivars in the survey, the number of distinct cultivars represents the local stock of cultivar attributes. A positive relationship between banana area and cultivar demand is expected. No direction of effect is hypothesized for the relationship between a greater availability of different types of banana planting material in a village and the demand for any single cultivar. A greater range of choice may lead to more or fewer types grown per plantation.

Household transaction costs are represented by a measure of the time needed to get to the nearest banana market (Table 6.1) and the supply (or farm-gate) price (Table 6.2). The farm-gate price is calculated using a triangular distribution for bunch prices

[6] Experience is corrected by age to reflect population dynamics in Uganda, where high mortality rates among the active adult population cause young people to be relatively more experienced for a given task than would otherwise be the case. The banana production decisionmaker is not necessarily the household head, but the person in charge of banana production and management decisions in the household. In Uganda, this person is often a woman.

elicited from the farmer for the previous year, by cultivar (Hardaker, Huirne, and Anderson 1997). It is hypothesized that farmers have a greater incentive to grow more plants on their farms if it takes them longer to reach the market place (because of poor road quality or lack of adequate means of transportation), and if the banana bunch price they receive at the farm gate is higher.

Scientific evidence, as well as previous participatory research, was used to cull a subset of the most important attributes and traits considered by banana farmers when they choose banana cultivars (Gold et al. 1998). Farmers were asked to rate attributes according to their importance and each banana cultivar according to its possession of the attributes (adapted from Reed, Binks, and Ennew 1991). Consumption attributes include cooking quality, measured as a categorical variable. Expected yield losses were calculated from subjective yields elicited as triangular distributions (Hardaker, Huirne, and Anderson 1997) in the presence and absence of black Sigatoka and weevils, and observed incidence of the disease or pest. To aid farmers in recognizing the trait or attribute, colored photographs of each attribute of interest were presented. Physical differences in the effects of both pathogens are readily observable and visually distinguishable. Estimates calculated from subjective yield distributions compared broadly with parameters tabulated from existing on-farm trial data (Gold et al. 2004), although clearly, farmers' decisions are made on the basis of their own perceptions and experience. An inverse relationship is anticipated between cultivar demand and the expected yield loss variables.

Table 6.3 Zero-inflated Poisson (ZIP) regression results

Variable	Estimated coefficient	*p*-value
Constant	2.5113***	0.000
Experience	0.3590***	0.000
Education	0.0322***	0.000
Dependency ratio	0.1534***	0.000
Household size	0.0732***	0.000
Extension	0.0176***	0.000
Assets	- 0.0007***	0.000
Exogenous income	- 0.0002***	0.000
Banana area	0.0402***	0.000
Planting material	- 0.0492***	0.000
Low elevation	0.4030***	0.000
Time	0.0289**	0.059
Farm-gate price	0.0377***	0.001
Cooking quality	0.1701***	0.000
Loss due to black Sigatoka	- 0.0053***	0.000
Loss due to weevils	- 0.0012**	0.032

N = 1,148 (zero observations = 199)
Log likelihood: –11,859.53
Vuong test of ZIP versus Poisson: z = 9.73 (p-value = 0.000)
Note: *** and ** indicate statistical significance at the 1 percent and 5 percent levels, respectively, according to the hypothesized one-tailed or two-tailed tests.

Results

Determinants of Demand for Planting Material of Cooking Cultivars

Following the formulation of our empirical approach, from the 3,619 observations (517 households × 7 cultivars per household), 1,148 household-cultivar observations, belonging to the set $\tilde{v}$ were used to estimate cultivar demand. All explanatory variables are statistically significant for cultivar demand. The hypothesized directions of the effects are mostly consistent with expectations (Table 6.3).

Larger households and households with a higher proportion of dependents grow more banana plants to meet their immediate consumption requirements, and the size of the effect of the dependency ratio is relatively large. The positive association with experience and education implies that acquisition of human capital has an incremental impact on the extent of banana production. More frequent visits of extension agents appear to increase cultivar demand, perhaps due to better knowledge associated with each cultivar and related management practices. Wealthier households appear to have lower demand, suggesting that income effects drive substitution away from home production of cooking bananas.

Banana area is positively related to cultivar demand, as expected. Greater availability of different planting materials reduces demand for any single cultivar. This finding likely reflects complementarities in the bundles of consumption and production attributes provided by different banana cultivars, which motivates farmers to plant smaller numbers of more cultivars on the available land. Cultivar demand is also greater for households in low-elevation areas, the historical locus of banana production in Uganda, by a large magnitude.

Households located farther from markets have greater cultivar demand, meeting their immediate household consumption needs through own production. A positive supply response is evident, however, with a higher farm-gate price inducing farmers to plant more mats and hence sell more banana bunches. Because all households sell bananas at the farm gate, the variable "distance to market" captures the behavior of buyers. The farm-gate price is related to the behavior of sellers.[7]

The statistical importance of consumption and production attributes in understanding cultivar demand is clear. Perceived yield losses caused by either black Sigatoka or weevils reduce cultivar demand significantly, evidence that farmer demand for planting material is responsive to improving resistance through effective gene insertion for targeted traits. Perceived cooking quality in a cooking-type cultivar raises demand significantly and by a large magnitude, highlighting the importance of consumption attributes when host-plant material is selected for gene insertion.

Sensitivity Analysis

Postestimation, we use the fitted model to illustrate several policy messages concerning the potential demand of Ugandan farmers for genetically engineered planting material ex ante. First, "client" prototypes are developed by comparing the characteristics of households at the upper and lower tails (20 percent) of the distribution of predicted demands. Industry clients are the households with the highest predicted demand for host planting material with the genetically engineered traits. Examining the mean characteristics of households located at the upper and lower tails of the distribution of predicted demands provides insights that are not readily apparent in the regression relationship, which depicts marginal changes on average. This exercise demonstrates that scientists' choice to transform one host cultivar rather than another can have social consequences, favoring one rural population over another. Such choices are often necessary, however, because of technical constraints and the costs of transformation, particularly in a vegetatively propagated crop, such as banana.

Second, the total farmer demand for planting material of a genetically transformed host cultivar is calculated as an indicator of the size of the banana industry demand. Given the vegetatively propagated nature of the crop, some public investments in the planting material industry will probably be necessary, perhaps in conjunction with private investments in biotechnology tools and the investments

[7] The bulky nature of banana bunches constrains their transportation to local trading centers or urban markets. Thus, the point of sale is generally the farm gate. In our survey, only a small fraction of the farmers also sold bunches at local trading centers. At the farm gate, transactions costs are typically borne by buyers (middlemen, other farmers), and they are reflected in the level of farm-gate prices received by selling households.

by local farmers' associations (de Vries and Tonniessen 2001). This exercise shows how a model of this type might be useful in assessing the investments needed to support the systematic dissemination of improved planting material.

Third, the sensitivity of farmer demand for improved planting material is demonstrated for several public investment scenarios. In the first scenario, we consider different degrees of effectiveness in gene insertion and expression. In the second scenario, we model the insertion of multiple compared to single traits. In the third scenario, we add supporting investments in infrastructure, education, and extension. The first two investments are related to technology development or investments in agricultural R&D. The third represents investments that can facilitate technology diffusion among farmers. Economists who study agricultural development have extensively documented the impacts of public investments in agricultural R&D, education, extension, and infrastructure on the rate of growth in agricultural productivity. Although these investments can compete for scarce public resources, they are often complementary in their effects on total productivity (Pingali and Heisey 2001; Evenson, Pingali, and Schultz 2007). Here, we use the sensitivity analysis to explore similar relationships in farm-level data.

Client Prototypes

First, the prototypes of likely adopters were compared across the seven cooking cultivars, using the upper tails of the distribution of predicted demand for planting material for each cultivar. Households most likely to adopt each one of the host cultivars share similar characteristics. Then, the prototypes of likely adopters were compared to those less likely to adopt the host cultivars by using the upper and lower tails of the distribution of predicted demand for planting material. Significant differences were identified between the prototypes of likely adopters and nonadopters. Table 6.4 compares the characteristics of households identified in the upper and lower tails of this distribution for the most widely grown cooking cultivar, Nakitembe, as an example.

Differences between the characteristics of the population segments with high and low potential demands indicate that the choice of cultivar for genetic transformation is likely to have social implications. For example, households with high potential demand for transgenic Nakitembe are larger, with more dependents and less-educated members. They are poorer in cash transfers and livestock assets, and less likely to live in a brick house. They manage larger farms, with a lower share of land allocated to banana groves. They tend to spend more time reaching banana markets even though they are not more distant from them, perhaps because of poor transport or road quality. They are located in the Central Region of the country, where banana production is most constrained by biotic and population pressures. This region is also the primary destination of banana output in the country, with the highest concentration of urban consumers of cooking bananas.

The Size of Demand for Planting Material

Total population demand for a potential host cultivar is calculated as the product of the weighted total predicted demands in the sample and the proportion of banana-growing households in the domain:

$$(h = 1,\ldots,H)\colon \left(\sum_{h=1}^{H} \Psi_h \hat{v}_h\right) \frac{H}{540}.$$

The weights (Ψ) are calculated using sampling fractions defined in the sample design, and 540 is the total number of households in the stratified sample. Again, we use the widely grown cooking cultivar Nakitembe as an illustration. The baseline total demand for this cultivar in the study domain, assuming no resistance to biotic pressures and no supporting investments, is around 250,000 banana mats, as shown in Figure 6.1.

Table 6.4 Prototype households with high and low predicted demands for the potential host variety Nakitembe

	High demand (N = 72)	Low demand (N = 72)
Proportion of male decisionmakers (percent)	62.39***	46.06***
Experience of decisionmaker (years)	9.94**	7.37**
Household size	7.77***	4.10***
Proportion of dependents in household (percent)	67.31***	40.60***
Proportion of women in household	51.61	46.54
Average education of females in household (years)	3.28*	3.76*
Average education of males in household (years)	4.12***	5.42***
Assets (livestock value) (thousand Ush)	33.28***	85.69***
Exogenous income (thousand Ush)	52.25***	364.33***
Proportion of households living in brick house (percent)	57.99*	59.01*
Farm area (acres)	5.75***	3.46***
Banana area (acres)	1.04**	0.73**
Banana share	0.34***	0.41***
Time to market (hours)	1.37***	0.87***
Distance to market (kilometers)	1.87	1.62
Proportion of net selling households (percent)	63.00	45.47
Proportion of net buying households (percent)	35.26	53.83
Proportion of households in Southwest Region (percent)	3.25***	18.47***
Proportion of households in Central Region (percent)	81.34***	70.57***
Proportion of households in Eastern Region (percent)	15.42***	10.96***

Note: ***, **, and * indicate statistically significant differences between means and/or distributions at the 1, 5, and 10 percent levels, respectively.

Figure 6.1 Change in predicted total industry demand for planting material of Nakitembe from effective trait insertion and other supporting investments, for different levels of resistance, by scenario

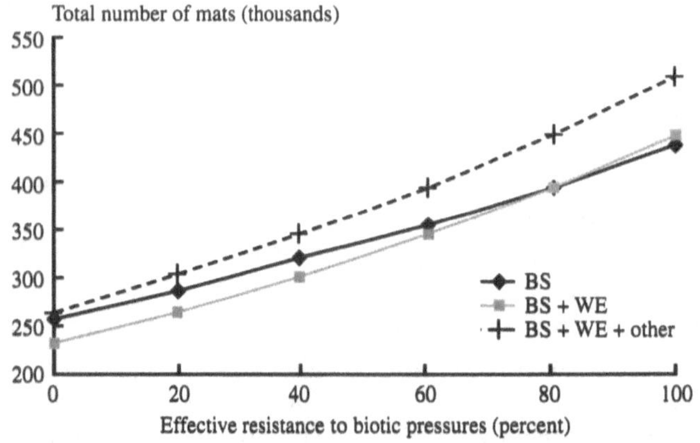

Notes: BS is black Sigatoka and WE is weevils. The intercepts (in thousands) for the three lines are 259 (BS), 233 (BS + WE), and 267 (BS + WE + other). Their slopes (in thousands) are 36.01 (BS), 42.38 (BS + WE), and 48.55 (BS + WE + other).

Sensitivity of Farmer Demand

The fitted model is used to analyze the sensitivity of total demand for host cultivars to genetic transformation by varying the perceived yield losses due to weevils and black Sigatoka fungal disease. Biotic constraints can be considered individually and jointly, demonstrating the gains that might be achieved in use levels through multiple compared to single gene insertion. Next, the effects of removing impediments to use through other public investments in education, market infrastructure, and extension are depicted, given the research investment made in genetic transformation. An increase in the supply (or farm-gate) price from a mean value of Ush 2,000 to Ush 4,000 is assumed as a result of improvement in bunch size and quality or market efficiency. The time taken to get to a nearby banana market is halved from a mean value of 1 hour, reflecting improvements in roads or better means of transportation. The education of the banana production decision-maker is increased through completion of primary school (7 years) from a mean value of 5 years. Finally, more visits by extension agents are made to improve dissemination of information and knowledge about banana production, including understanding of pests and diseases.

Nakitembe is again used as a representative example. Figure 6.1 depicts the sensitivity of farmer demand for planting material to changes in the effective resistance to biotic pressures when a single resistance gene (for black Sigatoka) or multiple resistance genes (for black Sigatoka and weevils) are successfully inserted into suitable banana genotypes, including the effect of supporting public investments. The demand increases twofold when full resistance is achieved, and diffusion of improved planting material is reinforced through investments in education, market infrastructure, and extension.

Sensitivity of cultivar demand to insertion of multiple resistance genes is greater than the more gradual effect of reduction in loss from only one biotic constraint. For a crop affected by a complex of biotic pressures, the marginal effect on expected demand will be much greater (close to a 50 percent increase in the total number of plants) when insertion and expression of more than one resistance trait are jointly achieved.

Note that two fundamental conditions are necessary for outcomes such as these to be borne out in the banana groves of Uganda, and neither of them is sufficient. The first condition, which is technological, is that genes must be successfully inserted and expressed. The second concerns farmers' perceptions. Even if the gene is inserted and expressed, farmers may not perceive its effects, because disease or pest incidence is variable across a banana plantation, and the mechanisms generating some effects (such as weevils) are better understood than others (such as black Sigatoka, a more recently introduced disease). Where necessary, perceptions and understanding of losses from pests and plant diseases can be enhanced through educating farmers.

As expected, overcoming impediments to adoption through public investments in education, extension, and market infrastructure augments the magnitude of the effect on demand induced through scientific (genetic) innovations. The combined changes in these variables act more as a shift than as a pivotal adjustment of the total aggregate demand for the planting material. About 50 percent of the shift is attributed to farm-gate price response, with education and extension having a smaller, though important, influence on its magnitude. Shortening the time taken to reach a nearby market has an offsetting effect (though only by 2 percent) by enabling some substitution of market purchase for on-farm production of food. Improvements in market signals, such that quality differentials across banana bunches are captured, could stimulate price responsiveness, with important implications for

Table 6.5 Characteristics of households with high predicted demand for the potential host cultivar Nakitembe, before and after effective insertion of resistance to black Sigatoka and weevils, with supporting public investments

	High demand for Nakitembe (N = 72)		
Characteristic	Original distribution	60 percent resistance to BS	60 percent resistance to BS and WE plus other public investments
Percentage of decisionmakers who are men	62.39	61.67	48.01
Mean years of experience as a decisionmaker	9.94	10.96	9.79
Mean number of household members	7.77	8.10	7.42
Mean percentage of household members who are dependents	67.31	68.08	66.37
Mean percentage of household members who are women	51.61	53.54	50.00
Mean years of education of women in household	3.28	3.23	2.89
Mean years of education of men in household	4.12	4.27	3.56
Mean value of livestock (thousand Ush)	33.28	42.35	33.60
Mean income received, previous year (thousand Ush)	52.25	60.61	71.22
Percentage of households living in brick house	57.99	61.01	50.27
Mean farm area (acres)	5.75	6.04	5.14
Mean banana area (acres)	1.04	1.07	0.83
Mean percent of farmer area in bananas	34.	34.	34.
Mean hours to market	1.37	1.42	1.41
Mean distance to market (km)	1.87	1.96	1.71
Percentage of households that are net sellers of bananas	63.00	63.01	43.67*
Percentage of households that are net buyers of bananas	35.26	39.41	52.45*
Percentage of households in Southwest Region	3.25	2.72	1.13
Percentage of households in Central Region	81.34	82.10	82.63
Percentage of households in Eastern Region	15.42	15.18	16.25

Notes: * indicates statistically significant differences between means and/or distributions at the 10 percent level. The three columns are (1) the original distribution of predicted values; (2) the distribution of predicted values with 60 percent effective resistance to black Sigatoka (BS); and (3) the distribution of predicted values with 60 percent effective resistance to BS and weevils (WE), and support for public investments. Investments are described in the text. See footnote 9, Chapter 5, for an explanation of measurement of yield losses.

rural development. Improvements in road infrastructure or means of transportation appear to have a smaller effect on the sensitivity of total aggregate demand in the short term.

Changes in clients before and after effective gene insertion, with and without supporting public investments, are presented in Table 6.5. As expected, no substantive differences in the prototype households are identified when gene insertion alone is undertaken. When gene insertion is accompanied by other supporting investments, differences in market participation emerge, with a lower proportion of households participating as net sellers and higher proportion engaging in market transactions as net buyers. This finding may reflect either cultivar specialization or reductions in transactions costs that lead households to cross the threshold of autarky. Overall, changes in prospective clients suggest that enhanced resistance to biotic constraints is likely to have an impact on distribution of banana products only when accompanied by the improvement of market-related infrastructure (for example, roads).

Conclusions

The theoretical framework of the agricultural household is appropriate for studying cultivar demand decisions in Sub-Saharan Africa, because it takes into consideration imperfections in input and output markets. The trait-based adoption model presented here is particularly well suited to analysis of gene-insertion technologies compared to the earlier generation of improved cultivars bred through crossing. With these technologies, genes are inserted into a host cultivar without affecting the levels of other traits and attributes. Host cultivars can be modern cultivars bred by scientists or landraces maintained by farmers. Some crops are difficult to improve through crossing, as is the case for EAHBs. A key advantage of genetic transformation in banana improvement is that it is possible to introduce desirable traits without changing the end product—something that is very difficult to accomplish through conventional breeding.

Postestimation analyses are then used to illustrate several policy messages. First, comparisons of household prototypes with high and low predicted demand for potential host cultivars indicate that the clients for transgenic planting material are likely to be poorer, subsistence-oriented farmers in areas greatly affected by biotic constraints. Hence, this technology is potentially pro-poor. The choice of host cultivar for genetic transformation will probably have social consequences, and there are also direct costs associated with transforming multiple banana cultivars. Second, for biotic pressures and complexes, such as those considered here, insertion of multiple traits could have large benefits in terms of farmer demand for improved planting material, although these advantages must be weighed against the cost of transformation and regulatory aspects. Third, success in technology generation is a necessary but not a sufficient condition for its widespread use. Past experience has amply demonstrated that other factors can impede the adoption of even the most promising scientific innovations. The approach used here illustrates the potential effect of effective gene insertion but also the role that other public investments in education, extension, and infrastructure can play in supporting farmer demand for the new technology.

In general, targeting traits that reflect local production conditions and consumption preferences and identifying endemic banana cultivars as host plants may lead to greater acceptability of genetically engineered banana cultivars. The trait-based, household model of demand for genetically engineered planting material is amenable to different crops and settings. Although beyond the scope of the farm-level research analyzed here, other important aspects of introducing genetically engineered food crops in developing economies clearly merit investigation. These include the appropriate design of biosafety regulatory frameworks, consumer attitudes toward biosafety risk, and potential challenges of marketing transgenic products in domestic and foreign markets.

Genetic transformation is only one of several strategies to improve banana productivity, and one that depends heavily on other factors, including relieving technical constraints to transformation and establishing an appropriate biosafety framework. The next three chapters investigate two other strategies: the adoption of recommended management practices (Chapters 7 and 8) and the adoption of banana hybrids (Chapter 9).

References

Cameron, A., and P. Trivedi. 1998. *Regression analysis of count data.* Econometric Society Monograph 30. Cambridge: Cambridge University Press.

de Janvry, A., M. Fafchamps, and E. Sadoulet. 1991. Peasant household behaviour with missing markets: Some paradoxes explained. *Economic Journal* 101 (November): 1400–1417.

de Vries, J., and G. Toenniessen. 2001. *Securing the harvest: Biotechnology, breeding and seed systems for African crops.* Wallingford, U.K.: CAB International.

Edmeades, S., 2003. Cultivar choice and attribute trade-offs within the framework of agricultural household models: The case of bananas in Uganda. Ph.D. dissertation, North Carolina State University, Raleigh.

Englin, J., P. Boxall, and D. Watson. 1998. Modeling recreation demand in a Poisson system of equations: An analysis of the impact of international exchange rates. *American Journal of Agricultural Economics* 80 (2): 255–263.

Evenson, R. E., P. L. Pingali, and T. P. Schultz, eds. 2007. *Handbook of agricultural economics, 3A. Agricultural development: Farmers, farm production and farm markets.* Amsterdam: Elsevier.

Gold, C. S., A. Kiggundu, D. Karamura, and A. Abera. 1998. Diversity, distribution and selection criteria of *Musa* germplasm in Uganda. In *Bananas and food security,* ed. C. Picq, E. Fouré, and E. A. Frison. International symposium on the socioeconomics of non-export banana production, Douala, Cameroon, November 10–14, 1988. Montpellier, France: International Network for the Improvement of Banana and Plantains.

Gold, C. S., G. Kagezi, G. Night, and P. Ragama. 2004. The effects of banana weevil damage on highland banana growth, yield and stand duration in Uganda. *Annals of Applied Biology* 145 (3): 263–269.

Hardaker, J., R. Huirne, and J. Anderson. 1997. *Coping with risk in agriculture.* Wallingford, U.K.: CAB International.

Karamura, D., and B. Pickersgill. 1999. A classification of the clones of East African highland bananas (*Musa*) found in Uganda. *Plant Genetic Resources Newsletter* 119: 1–6.

Ladd, G., and M. Martin. 1976. Prices and demands for input characteristics. *American Journal of Agricultural Economics* 58 (1): 21–30.

Ladd, G., and V. Suvannunt. 1976. A model of consumer goods characteristics. *American Journal of Agricultural Economics* 58 (3): 504–510.

Lancaster, K., 1966. A new approach to consumer theory. *Journal of Political Economy* 74 (2): 132–157.

Pingali, P. L., and P. W. Heisey. 2001. Cereal crop productivity in developing countries: Past trends and future prospects. In *Agricultural science policy: Changing global agendas,* ed. J. M. Alston, P. G. Pardey, and M. Taylor. Baltimore and London: International Food Policy Research Institute/ Johns Hopkins University Press.

Reed, G., M. Binks, and C. Ennew. 1991. Matching the characteristics of a service to the preferences of customers. *Managerial and Decisions Economics* 12: 231–240.

Singh, I., L. Squire, and J. Strauss, eds. 1986. *Agricultural household models: Extensions, applications, and policy.* Baltimore: Johns Hopkins University Press.

Tripp, R., 2001. Can biotechnology reach the poor? The adequacy of information and seed delivery. *Food Policy* 26 (3): 249–264.

von Haefen, R., D. Phaneuf, and G. Parsons. 2004. Estimation and welfare analysis with large demand systems. *Journal of Business and Economic Statistics* 22 (2): 194–205.

Vuong, Q. 1989. Likelihood ratio tests for model selection and non-nested hypotheses. *Econometrica* 57 (2): 307–334.

CHAPTER 7

Social Capital and Soil Fertility Management in the Banana Production Systems of Uganda

Enid Katungi, Melinda Smale, Charles Machethe, and Wilberforce Kateera Tushemereirwe

The research in this chapter was motivated by the hypothesis that long-term improvement in banana production in Uganda will be achieved by simultaneously solving the problems of pests or diseases and soil fertility depletion. Despite the recognition of its importance, comparatively less research has been directed toward the problem of soil fertility depletion. Information about the factors that influence whether banana farmers invest in soil fertility management is scant. The underlying conceptual approach in this chapter is based on the model of the agricultural household described in Chapter 2. A factor analysis is used to cluster a set of soil fertility management practices that are jointly applied into those related to mulch and manure application. The influence of household, farm, market, and social characteristics of farmers on decisions about soil fertility management is estimated using a probit regression, an ordinary least squares regression, and a two-step Heckman estimation procedure. The analysis assesses the extent to which recommended practices are known and used by farmers; identifies the determinants of use; and draws implications for research, extension, and national policy. The data and sample are described in Chapter 2 and Appendix D.

Theoretical Model

The model presented here depicts how a semisubsistence banana grower in Uganda may choose to allocate banana area between a soil fertility management technology that enhances the productivity of land but is labor intensive and a traditional technology for managing soil fertility. The basic agricultural household model described in Chapter 2 is extended to incorporate social capital as a component of exogenous income and information accumulation processes.

Consider a household that both produces and consumes bananas, buying inputs and selling produce in imperfect markets. Banana can be produced using two alternative soil fertility management technologies: the improved management technology (f^I) and the traditional management technology (f^T). The improved management technology utilizes labor (L) and two types of external organic fertilizers (fertilizers in the form of mulch and manure application) expressed in the vector F. Organic fertilizers are mainly produced on the farm as by-products of other farm activities, and there is virtually no market for sales or purchase of these inputs.[1] Use of the organic fertilizers increases the productivity of land allocated to banana, which increases the amount of bananas available for household consumption and the surplus for sale. The yield effect of organic fertilizers is significantly superior to those of the traditional technology, especially when soil fertility has deteriorated. When markets are imperfect, differences in household endowments create differences in levels of investment in soil management practices.

The application of organic fertilizers requires additional resources in the form of labor or cash income to hire labor. The farmer incurs some variable cost (in terms of time or money paid to hired labor) to gather, transport, and apply the organic fertilizers. Reliance on family labor in production implies that leisure is valued by its marginal worth to the household rather than as an opportunity cost imputed from a market wage rate.

The farmer also bears a fixed cost of information acquisition. Information about the recommended technology can be obtained from extension educators, other farmers' experiments in the community, and/or from own experimentation. Following the structural adjustment programs implemented in Sub-Saharan Africa, including Uganda, access to agricultural extension by the majority of small farmers was curtailed significantly. Farmer experimentation and social learning are currently the primary sources of information about new technologies. Even when extension is provided, agents tend to give general information, and a farmer who is less familiar with the improved technology may decide to allocate a small portion of his/her land to gain more knowledge about its relevance, resulting in partial use of the technology.

Banana output under the improved management technology (f^I) is:

$$q^I = f^I\left(A^I, L, F \mid \Omega_F, \Omega_M, K_t\right). \qquad (1)$$

Following the model of Isham (2000) and Feder and Slade (1984), it is assumed that the stock of knowledge evolves as a function of experience in years with the technology (τ), a set of diffusion parameters (Ω_D) and different forms of social capital (Ω_S). Diffusion parameters include cumulative contact with the extension educators and other formal information institutional factors within the community. The stock of knowledge can be expressed as:

$$K_t = k(\tau, \Omega_D, \Omega_S).$$

The traditional management technology uses only labor (L) and land to produce banana output for a given set of farm characteristics (Ω_F) and market constraints (Ω_M). The knowledge stock is excluded from the banana production using the traditional management technology, because the technology has been in the communities for many years and all farmers are assumed to have full information about it. Banana output under the traditional management technology is given by the following production function:

$$q^T = f^T\left(A^T, L \mid \Omega_F, \Omega_M\right). \qquad (2)$$

The two technologies compete for land allocated to banana production ($A^I + A^T = \bar{A}$). The household can choose to manage all its bananas with the improved management technology or with the traditional management technology. The household also has the option to allocate part of the banana area under the improved management technology and the remainder of the banana area to the traditional technology. The share of the banana area the farmer allocates to the improved management technology (δ)

[1] Household demand for this type of input is conditional on its own supply.

ranges from 0 to 1. Given a pre-allocated area in banana production ($\bar{A}$), total banana output obtained by the farmer is:

$$Q^B = f^I\begin{pmatrix} A', L, F \mid \boldsymbol{\Omega}_F, \boldsymbol{\Omega}_M, k(\tau, \boldsymbol{\Omega}_D, \boldsymbol{\Omega}_S) \\ +f^T(\bar{A} - A', L \mid \boldsymbol{\Omega}_F, \boldsymbol{\Omega}_M) \end{pmatrix}. \quad (3)$$

The household chooses levels of inputs and consumption of banana (X^B) and purchased goods (X^G) to maximize the expected utility of consumption and leisure (h), given its characteristics ($\boldsymbol{\Omega}_{HH}$) and the characteristics of the market ($\boldsymbol{\Omega}_M$):

$$\max_{\Psi} U(X^B, X^G, h \mid \boldsymbol{\Omega}_{HH}, \boldsymbol{\Omega}_M). \quad (4)$$

Household characteristics, such as the gender and age of the banana production decisionmaker, the size of the household, income, landholdings, and livestock assets, influence the marginal utilities of the consumption items and hence the consumption preferences of the household. Markets are missing for organic fertilizers, and the household cannot demand more of these than it can supply from its own sources.

The household also faces a time constraint. Time allocated to banana production (L), and leisure (h) cannot exceed the total stock of time owned by the household ($T \geq L + h$).

Utility is maximized subject to a full income constraint:

$$P^B(Q^B - X^B) + e(I, \boldsymbol{\Omega}_S) = P^G X^G. \quad (5)$$

Full income is the value of the marketed surplus (P^B $(Q^B - X^B)$ and exogenous income (e). The constraint excludes the time endowment, because its opportunity cost is endogenous. Under imperfect market conditions, the only income to the household at the time of making decisions consists of cash received through private assets or social capital. Represented by the function $e(I, \boldsymbol{\Omega}_S)$, exogenous income consists of rent or interest earned privately (I) and/or in the form of bilateral transfers, such as free labor, gifts, remittances, or informal credit from social networks ($\boldsymbol{\Omega}_s$). Household income is spent on purchasing other goods (X^G) consumed by the household at market price (P^G).

The farmer has a prior belief (θ) that the soil fertility problem exists, with a belief ($1 - \theta$) that it does not. The farmer's belief is based on his experience with the banana production, knowledge stock, and physical land characteristics. Conditional on the deterioration of the soil fertility, the yield gains are higher under the improved management technology than the traditional management technology. The yield gain from the improved soil fertility management practices is indeterminate when there is no soil fertility problem. The yield gains can also be zero given the fixed genetic yield potential of the crop.

Kuhn-Tucker conditions are used to derive optimal choices of the crop management technology. Given that conditions for an optimum are met, the following structural equation defines the optimal share of banana area managed with the improved technology:

$$\begin{aligned} \delta^* = \delta(P^B, \boldsymbol{\Omega}_{HH}, \boldsymbol{\Omega}_F, \boldsymbol{\Omega}_M, e(I, \boldsymbol{\Omega}_S); \quad (6) \\ k(\tau, \boldsymbol{\Omega}_D, \boldsymbol{\Omega}_S); \Theta(\boldsymbol{\Omega}_F; k(\tau, \boldsymbol{\Omega}_D, \boldsymbol{\Omega}_S, \boldsymbol{\Omega}_F,))) \end{aligned}$$

The optimal share (δ^*) is bounded from below at 0 and from above at 1. Equation (6), the basis of the econometric estimation, embeds the hypothesized conceptual determinants of knowledge accumulation and formation of perceptions processes, as distinct from factors that influence both knowledge about practices and their use.

Econometric Estimation

The econometric analysis in this chapter focuses on the household demand for mulching and manure, both divisible technologies. The econometric approach can be linked to the theoretical model through an index function model involving decisions about whether to use the technology and how much of it to use. Denote by $\mathbf{y}^*$ a vector of the unobserved demand for the improved soil fertility management technology. The

Table 7.1 Results of the likelihood ratio test for the null hypothesis that the coefficients on the two management decisions are equal

	Value of the log likelihood function			
Management practice	Tobit model	Probit model	Truncated regression	Likelihood ratio test p-value
Mulching	–90.358	–147.68	134.69	0.000
Manure	–110.90	–163.50	91.90	0.000

general structure of the demand equation can be expressed as:

$$y^* = \boldsymbol{\alpha}' \mathbf{Z} + \boldsymbol{\mu} \quad \boldsymbol{\mu} \mid \mathbf{Z} \sim N(0,1). \qquad (7)$$

Z represents a vector of explanatory variables cast on the right hand side of equation (6), **μ** is a vector of unobserved heterogeneity and **α** is a vector of parameters to be estimated. At the time of the survey, demand was not observed for some households, and hence we cannot estimate the structural equation. Instead we estimate a reduced form and focus on two management decisions, the discrete decision (to use or not to use) and the extent of use.

Several problems were encountered when estimating the demand of soil fertility management technology. First, the extent of use decision (δ^*) is observed only when the outcome from the decision to use or not to use the technology is positive. Missing data for a set of explanatory variables lead to a censoring of the demand for the management practices for which the discrete decision is nonpositive (Maddala 1983; Greene 2000; Wooldridge 2002).

A tobit regression model has been widely used in estimation when the dependent variable is observed within a limited range (Greene 2000). Underlying the tobit model is the assumption that the coefficients on the probability and extent of adoption are the same (Greene 2000). Thus, the tobit model fails to separate the two decisions that characterize the adoption of a divisible technology. The decision to use and the extent of use are also likely to be influenced by different factors (Wooldridge 2002).

To test whether the tobit model is a suitable representation of the processes affecting the use of mulching or manure technology, probit, tobit, and truncated regressions were estimated for each of the two technologies. The null hypothesis of equal coefficients was tested using the likelihood ratio statistic, where the restricted regression is the tobit model and the unrestricted regression is the combined probit and truncated regression. Results are summarized in Table 7.1. For both manuring and mulching, the statistical significance of the test-statistic leads to rejection of the null hypothesis that the coefficients are equal. The data therefore support separate estimation of the use and extent of use decisions.

The reduced-form model describing the share of banana area allocated to mulching or manure technology in the population is:

$$\begin{aligned} \boldsymbol{\delta}^* \mid y = 1 &= \boldsymbol{\beta}'\mathbf{X} + \boldsymbol{\varepsilon} \\ y = 1 &\quad \text{if} \quad y^* > 0 \\ y = 0 &\quad \text{otherwise.} \end{aligned} \qquad (8)$$

As defined above, $\boldsymbol{\delta}^*$ is the optimal observed area share of the bananas under the technology, **X** is a vector of explanatory variables, and $E(\boldsymbol{\varepsilon} \mid \mathbf{X}) = 0$; it is assumed that the unobserved heterogeneity in the vector **ε** is uncorrelated with the exogenous variables.

The second problem is that the soil fertility management decisions are also conditioned on the farmers' perceptions of the soil fertility problems, a random variable. Farmers' perceptions about the soil fertility problem are influenced by the same variables that influence the use of mulching or manure technology. Thus, we are dealing with a simultaneous equations model that is a function of exogenous variables, predetermined variables, and an error term. Banana management decisions can be estimated as a function of direct measures of perception or by substituting for perception using exogenous factors in the perception equation. Use of the direct measure of perception in the management equation creates the problem of endogeneity. The observed indicator of perception is correlated with the error term of the demand equation, rendering the ordinary least squares estimates inconsistent (Greene 2000; Woodridge 2002). Consistent estimates can be obtained by use of a two-stage least squares estimator to correct for endogeneity (Wooldridge 2002). If correctly specified, this estimation procedure will yield estimators with greater asymptotic efficiency than attainable by the limited information method (Greene 2000). However, the approach requires extensive data, which in most cases are not available. In addition, the full information method is complex when the null hypothesis of sample selection bias has not been rejected. Because our main interest in this chapter is to test the effect of social capital while controlling for other factors, demand for soil fertility management is estimated as a function of all exogenous variables cast on the right hand side of equation (6), expressed as a reduced form.

The extent of use of mulching and manure application was estimated using two methods: ordinary least squares regression and the two-step Heckman model. The latter model accounts for the selection bias associated with missing observations for a given subsample caused by the truncated nature of the dependent variable. The motivation underlying the use of either the ordinary least squares or Heckman regression model is based on their statistical performance according to whether the null hypothesis of sample selection bias was rejected or not. A two-stage Heckman model was also used to test for sample selection bias. Although the share ranges from 0 to 1, the sample data indicate that all households have an extent of use that is less than 1. Thus, a model that accounts only for censoring at 0 was applied.

The probability of using mulch or manure was estimated with a probit regression in the first stage of a Heckman procedure using the whole sample. The inverse Mills ratio, computed from the probit regression, was included in the second stage to test for selection bias (Greene 2000; Wooldridge 2002). Hypothesis tests do not support the presence of sample selection bias in the extent of use of mulching practices, but support it in the case of manure. Test results imply that in the mulching equation, the subsample of household with nonzero use is representative of the population (Wooldridge 2002). Consequently, the extent of use of mulching was estimated by ordinary least squares regression on a subsample with nonzero use, while a Heckman model was used to estimate the extent of manure application.

Although the explanatory variables in the first- and second-stage regressions are identical, nonlinearity of the inverse Mills ratio allows the identification condition to be met (Wooldridge 2002). The problem of multicollinearity, often induced by use of the Heckman procedure, was tested using the variance inflation factor (VIF) technique in STATA© 8.0 (StataCorp, College Station, Texas). All explanatory variables included in the estimation had VIF less than 5.0, which suggests that multicollinearity does not affect the results.[2]

Finally, when the factors that influence demand of one technology are also likely to influence demand for other technologies, it is possible that the unobserved heterogene-

[2] A VIF greater than 10 indicates a collinearity problem (Kennedy 1985).

ity in the two technology demand equations is also correlated. Estimating them jointly was not considered to be worthwhile, because the set of explanatory variables is identical in both equations (Greene 2000, 616), implying no gains in statistical efficiency. Thus, each technology was treated separately.

Dependent Variables

We considered eight practices recommended to farmers for soil fertility management in banana groves: (1) mulch with grass; (2) mulch with crop residues; (3) mulch with kitchen residues; (4) addition of cattle manure, (5) goat manure, (6) pig manure, or (7) poultry manure; and (8) composting of homestead refuse to make manure (compost manure). The eight practices were subjected to a factor analysis procedure using principal components to analyze correlation between them and determine whether they could be represented by a smaller number of components.[3] Based on the criterion of an eigenvalue greater than unity, the eight soil management practices were grouped into four independent packages according to four unobserved factors.[4] The latent factors were interpreted as mulching, animal manure, composting, and traditional technologies.

Based on the results from the factor analysis, the eight types of organic fertilizers were collapsed into two dependent variables: mulching practices and manure application. Use of mulching technology is defined as the practice of applying crop residues (nonbanana crop residues) or grass to mulch banana plantations. Use of this type of mulch materials in banana production involves costs for gathering, transportation, and application, which may not be affordable by some households. Use of manure technology is defined as the practice of applying animal waste (from cattle and pigs) or compost manure in the banana plantations. These organic fertilizers also involve a cost of access, transportation, and application that may limit their use.

The omitted category $(1 - \delta)$ is the traditional method of recycling banana residues and applying household refuse, used by virtually every farmer on at least some mats. Organic household refuse (kitchen residues, goat and poultry waste) is collected in mixtures from the homestead as part of the cleaning activity and spread between mats to control weeds, though some farmers reported that use was not deliberate. Farmers are advised to compost the household refuse and other organic materials before applying it in banana plantations to facilitate quick decomposition (Tushemereirwe et al. 2003). Therefore, this method may not be adequate for soil fertility management in banana plantations. In addition, little additional information would be gained by investigating

---

[3] A principal components analysis with Varimax rotation was applied (STATA 8.0). Principal components analysis partitions the sum of the variances for a set of variables by taking uncorrelated, linear combinations of those that account for the latest eigenvalues of the covariance matrix. The components are not easily interpreted, however. The Varimax rotation procedure, one of several types of rotation, enables a more straightforward interpretation of components in terms of factors "loading" high on some variables and low on others (Stevens 2002).

[4] The four factors explain 73 percent of total variance in the adoption components. Only variables with factor loading equal to or greater than 0.5 were considered significantly correlated with the underlying latent variable and hence grouped together. Factor 1 explains about 28 percent of the variance in the eight management practices. Organic household refuse (kitchen residues, goat manure, and poultry manure) were highly correlated with the first latent factor. Mulching techniques using the organic materials that are gathered from sources other than the banana crop (crop residues and grass) loaded on factor 2 and explain 18 percent of the total variance in the use of soil fertility management practices. Organic manure from animals that are rarely kept in the homestead (cattle and pigs) also clustered together, while the technique of composting household refuse before applying it to banana plants instead of applying it directly seems to be used independently of other soil fertility management practices.

the determinants of use of this technology, because almost every farmer uses it.

Explanatory Variables and Hypotheses

Table 7.2 shows the definitions of explanatory variables, summary statistics, and hypothesized effects. Conceptual variables included in the analysis of the soil fertility management decisions are given by reduced-form equation, and their operational definition was guided by the literature. Banana production is semisubsistence across much of the survey domain, with uneven access to markets and market participation, consistent with the nonseparable case of the household model. When the consumption and production decisions are nonseparable, comparative statics effects are ambiguous. Thus, hypothesized effects are motivated by related adoption literature and previous information concerning banana production in Uganda.

The structural equation (6) of the household model indicates that the share of banana area allocated to an improved soil fertility management practices is a function of village wage rate for unskilled labor (W), banana market price (P^B), household characteristics ($\mathbf{\Omega}_{HH}$), market characteristics ($\mathbf{\Omega}_M$), farm characteristics ($\mathbf{\Omega}_F$), exogenous income from private assets (I), formal information diffusion ($\mathbf{\Omega}_D$), farmer experience with the technology (τ), and different forms of social capital ($\mathbf{\Omega}_S$).

A number of household characteristics ($\mathbf{\Omega}_{HH}$) are hypothesized to directly or indirectly influence the use and extent of use of the mulching and manure technologies. The effect of these factors on management decisions depends on the nature of rural market imperfections (Pender and Kerr 1996). When labor markets are imperfect, households endowed with family labor may be able to meet the high labor demand of soil mulching and manure application compared to their counterparts with smaller endowments of family labor. Similarly, when markets are missing for organic fertilizers, households endowed with the assets (such as land and livestock capital) that produce these materials will be able to invest more in managing soil fertility. Boserup (1965) hypothesizes that increasing population pressure stimulates use of land intensification techniques. Thus, although more extensive landholdings per capita enable households to engage in crop production that may result in organic materials for mulching, it is associated with less pressure on the cultivable land. Low pressure on land may suppress perception of a soil fertility problem, reducing demand for soil fertility management practices.

The effects of owning livestock also appear to be ambiguous. Possession of livestock reduces the cost of access to animal manure, which may result in greater use. However, accumulation of livestock may imply a shift away from crop production and consequently reduced pressure on land, which may lower perceptions of a soil fertility problem and hence depress use of soil fertility management practices.

Education is another important household characteristic in adoption of technologies (Feder, Just, and Zilberman 1985; Feder and Umali 1993). Higher education is associated with the capacity to understand technical aspects related to soil degradation and conservation, but it may also increase the opportunity cost of labor that could reduce use of labor-intensive management practices. Older decisionmakers are expected to discount the future heavily, implying that age is associated with low investment in conservation practices (Shiferaw and Holden 1998).

Distortions in output market encourage self-sufficiency in that output, meaning that an increase in consumer-worker ratio (dependency ratio), which increases household consumption demand, will stimulate investment in production. However, when insurance and labor markets also fail, as is the case in rural Uganda, a higher consumer-worker ratio may increase the risk of starvation, which could limit the investment in labor-intensive activities. Hence the effect

Table 7.2 Definition of variables, hypothesized effects, and the summary statistics

Variable	Definition	Expected effect	Mean	Standard deviation
Use of mulch (δ_1)	Share of banana mats grown under dry grass or nonbanana crop residues		0.199	0.260
Use of manure (δ_2)	Share of banana mats grown under waste (from cattle or pigs) or composted manure		0.119	0.233
Farm characteristics ($\mathbf{\Omega}_F$)				
Elevation	Elevation at which the farm is located (1 = high, 0 = low)	+/–	0.281	0.450
Slope of the farm	Dummy equal to 1 if slope is rated steep and 0 otherwise	+	0.779	0.415
Soil moisture retention capacity	Dummy equal to 1 if soil moisture retention capacity of the banana plot is rated low and 0 otherwise	+	0.207	0.406
Drainage conditions in the plot	Dummy equal to 1 if the drainage conditions in the banana plot are rated poor and 0 otherwise	–	0.394	0.489
Mats	Total number of banana mat counts grown	+	283.43	334.677
Household characteristics ($\mathbf{\Omega}_{HH}$)				
Income (I)	Total income received by the household in form of rent and interest during the production year (000 Ush)	+/–	37.739	219.821
Age	Age of primary banana production decisionmaker (years)	–	43.076	15.670
Gender	Gender of the primary banana production decisionmaker (1 = male, 0 = female)	+/–	0.547	0.510
Education	Years of schooling of primary banana production decisionmaker	+	4.563	3.863
Household size	Number of people who live in the same homestead and eat from the same cooking pot	+	5.874	2.765
Land per capita	Total cultivable hectares of land (computed as total land possessed less area under forests, water/swamps, settlements) divided by the household size	+/–	0.372	0.677
Livestock unit	Sum of cattle units (0.8), sheep (0.4), pig (0.4), and goat (0.4) divided by the age of household head	+	0.032	0.042
Market characteristics ($\mathbf{\Omega}_M$)				
Price/wage ratio (P^B/w)	Average farm-gate selling price per kg of bananas in the village divided by the average wage rate for hired labor	+	0.095	0.033
Distance	Farm location from paved roads (km) as a measure of market access	+/–	10.689	7.050
Diffusion parameters ($\mathbf{\Omega}_D$)				
Extension	Number of cumulative contacts with the extension educators	+	1.151	2.049
Exposure	Dummy equal to 1 if a village was exposed to the new improved banana varieties and 0 otherwise	+	0.497	0.500
Experience mulch (τ_1)	Number of years mulching has been used in banana production on the farm divided by the age of the household head	+	0.251	0.244
Experience manure (τ_2)	Number of years manure has been used in banana production on the farm divided by the age of the household head	+	0.108	0.167

Table 7.2 (continued)

Variable	Definition	Expected effect	Mean	Standard deviation
Social capital (Ω_x)				
Membership density	Number of household members that belong to at least one association	+/–	1.251	0.975
Leadership heterogeneity	Continuous index measuring the degree of leaders heterogeneity in terms of livelihood or education higher than most group members in the village	+	4.441	0.559
Norms of decisionmaking	Continuous index measuring the degree to which decisionmaking in village associations is participatory	+	6.114	0.4725
Net labor transfers	Value (thousand Ush) of labor obtained from the social network less the value of labor supplied to the social network	+/–	0.339	8.153
Net cash transfers	Amount of cash (thousand Ush) obtained from its social network less the amount of cash supplied to the social network	+/–	–1.555	163.111
Net other item transfers	Value (thousand Ush) of other household items obtained from the social network less the value of those items supplied to the social network	+/–	0.296	39.791

of the dependency ratio on use of mulching and manure cannot be determined a priori.

The decision to invest in soil fertility management practices is also conditioned on farm characteristics (Ω_F), such as location, physical factors of the land, and scale of production. Elevation, a sample stratification parameter, is expected to condition both use of practices and perceptions. At higher elevations where there are more slopes on farms, the soil erosion potential is higher, which could stimulate farmer's perception of the soil fertility problem and consequently investment in soil conservation (Ervin and Ervin 1982). Prior biophysical information also indicates that banana productivity potential is higher in high-altitude areas, which provides further incentives to use soil fertility management practices. Physical characteristics of the land, such as the slope of the farm and soil moisture retention capacity of the banana plot, are hypothesized to directly affect perceptions of the soil deterioration. The slope of the farm represents the erosion potential while the capacity of the soil to retain moisture measures the capacity of the soil to support the high demand of the banana crop for water.

Therefore, low soil moisture retention capacity of banana plots should influence the demand for mulch and manure through its influence on perceptions. On the other hand, the counteracting effect of poor soil moisture retention capacity on the productivity of mulching or manure could lower the incentive to incur costly investments. Physical characteristics of land were measured in qualitative form using a subjective indicator reported by the farmers (Table 7.2). The drainage condition of the banana plot is another land characteristic important in banana production (Tushemereirwe et al. 2003). Bananas grown on poorly drained soils are more vulnerable to leaf spot diseases, which pose an exogenous risk in soil fertility management decisions.

The scale of production (measured by the banana mat count) reduces the fixed costs of information acquisition per unit area, thereby increasing the benefits from adoption (Feder and O'Mara 1981). The scale of production could also reduce the economic impact of soil depletion on the household, as does the landholding size. However, with respect to mulching technology, landholding size increases the access to mulch materials, while the scale of banana production may act, through fixed costs of information acquisition, to influence use. These mechanisms are conceptually different, justifying inclusion of both farm characteristics in the estimation.

The direct link between infrastructure development (Ω_M) and use of soil fertility management practices is not clear. Improvement in road infrastructure reduces the costs of physical access to markets, but also enhances market opportunities in nonagricultural enterprises, increasing the opportunity costs of investing in soil conservation. In the latter case, the indirect effects on use of soil conservation technologies might be negative when road infrastructure improves. Increased commercialization of banana production may also result in more soil depletion, which increases perception of the soil fertility problem and hence use of soil amendment techniques. Infrastructure development was represented in the estimation by distance from the center of the village to the nearest paved road, a village-level variable.

Economic theory predicts that better market price for bananas will increase the net returns from higher yields associated with soil fertility amendment technology, whereas higher input prices (for example, wage rate and the endogenous price of fertilizers) will reduce the returns and hence the incentives to use improved management technology. Because the banana farm-gate price is positively correlated with wage rate, a ratio of average farm-gate price of bananas to village wage rate was constructed for the estimation (P^B/w).

The expenditure constraint in the estimation is represented by the total exogenous income received by the household as net transfer in the form of interest from private assets, gifts, informal credit, and remittances. The exogenous income from household private assets (I) was defined as cash inflow in the form of rent from buildings or interest from previous investments. Households with access to this type of exogenous income are able to overcome liquidity constraints and can purchase farm implements that enable them to save on labor or use the cash to hire labor. Hence, they will be able to use and demand more of the soil fertility management practices.

Bilateral transfers from social networks also constitute an exogenous income to the household. Bilateral transfers in the form of gifts, informal credit, and labor are part of the major resources that rural households in developing countries exchange. These transfers can influence the demand for soil fertility management practices directly when used in implementing the practices or indirectly by reducing the risk aversion (Fafchamps and Lund 2003). Bilateral transfers accruing to the household were measured for 12 months by asking respondents whether the household had received any cash, any other items in the form of a gift, donations, or free labor from people other than household members in the past 12 months. For items received by the household, the respondent gave the total amount received; for noncash items, the concept of willingness to pay was used to attach a value to the items so as to standardize those items across households. Similar questions were asked about the household expenditure on each item in the social network for the same period. Net transfer for each item was computed as the difference between receipts and expenditure within a social network.

Information diffusion parameters are hypothesized to affect the use of practices both directly and indirectly through perception formation. Formal diffusion parameters included in the estimation are the cumulative contact with the extension educators and exposure to the new banana varieties. Large, discrete differences in knowledge are hypothesized between villages that have been exposed to new banana varieties and related information and those that have not. The positive role of extension educators in adoption of agricultural technologies is well established in the literature (Feder, Just, and Zilberman 1985; Feder and Umali 1993).

Social capital and farmers' experience represented informal means of generating or diffusing information. Experience with the technology affects both perceptions of soil fertility and use of soil fertility management practices. Social capital facilitates cooperation and willingness to share information by encouraging networking (Barr 2000; Isham 2000; Yli Renko, Autio, and Tontti 2002). Social capital variables included here are household density of membership in associations, norms of decisionmaking, and leader heterogeneity in associations. Household membership density is defined as the number of household members that belong to at least one association. Household membership density reflects the household's capacity to acquire information from the social network. However, the number of household members that join an association can also influence the opportunity cost of time used in banana production. Hence the nature of effect cannot be determined a priori.

Norms of decisionmaking and heterogeneity of leaders (in terms of livelihood and education) in associations were used as proxies for characteristics of associations. These proxies measure the group's ability to cooperate, share information, and network with other people beyond the village community.

Both norms of decisionmaking and leader heterogeneity are constructed at the village level. Norms of decisionmaking are defined as the degree to which members in an association participate in important issues of the association, computed as an additive index from responses to two questions regarding selection of leaders and decisionmaking of the associations. Respondents were asked about how important decisions are made. Responses were: 1 (only leaders participate), 2 (few members participate), and 3 (all members participate). They were also asked about how the leaders of each association to which they belong are selected. The responses were: 1 (by outside agents), 2 (each leader chooses a successor), 3 (by small groups of members), and 4 (by all members). The village index was computed by averaging over the number of responses in the village.

Leader heterogeneity is defined as the degree to which leaders in associations within a village differ from the rest of the people, computed as an additive index from responses to two questions concerning the wealth and educational status of each association leader. The responses were coded as: 1 (lower status than most members), 2 (same as most members), and 3 (higher than most members). The village index was computed by averaging over the number of responses in the village.

Results

Two sets of regressions results are presented in Table 7.3. The first set of regressions (probit models) estimate the effect of hypothesized factors on the decision to use or not to use mulch or manure. The second set of regressions predicts effect of the same hypothesized factors on the conditional demand for mulch and manure technologies while correcting for selection bias.

Decision to Use a Soil Fertility Management Practice

Marginal effects computed for the use decision in each probit model measure the expected change in the probability of observing a positive use of the technology with respect to a change in a particular explanatory variable at its mean value. In terms of overall percentage of predictions correctly classified, the model does well for both management practices, implying a good fit.

Applying the same equation structure (set of explanatory variables) for mulching and manure application reveals that the determinants of use are technology specific, with few patterns that can be discerned across technologies. Estimation results indicate that the physical characteristics of land (slope of the farm, soil moisture retention capacity, and drainage conditions in the banana plot) are important determinants of the decision to use soil fertility management practices, although the direction of the effect differs across technologies. At relatively higher slopes, the probability of using manure is low. This result contradicts the

Table 7.3 Factors influencing the probability and extent of use of soil fertility management practices

Variable	Probit model: Manure		Probit model: Mulch		Second step of Heckman model: Manure		Ordinary least squares model: Mulch	
	Marginal effects	Standard error	Marginal effects	Standard error	Coefficient	Standard error	Coefficient	Standard error
Farm characteristics (Ω_F)								
Elevation	−0.177	0.112	−0.042	0.110	−0.053	0.088	−0.116**	0.059
Plot drainage	0.137**	0.069	−0.008	0.055	0.037	0.045	−0.018	0.032
Moisture retention capacity	0.087	0.083	−0.121*	0.074	0.026	0.047	0.010	0.039
Slope of the farm	−0.144*	0.082	0.029	0.068	0.004	0.045	0.122***	0.036
Number of banana mats	8.3×10^{-5}	0.0001	0.0003**	0.0001	0.0001	0.000	−0.0001***	4.7×10^{-5}
Household characteristics (Ω_{HH})								
Age	−0.006***	0.002	0.0001	0.002	−0.002	0.001	0.0002	0.001
Gender	0.001	0.068	−0.021	0.055	0.012	0.041	0.050*	0.030
Education	0.004	0.009	0.013*	0.008	−0.002	0.006	0.011***	0.004
Household size	0.021	0.014	0.022***	0.012	0.020**	0.010	0.010	0.007
Dependency ratio	0.062	0.160	−0.218*	0.121	0.021	0.104	0.100	0.076
Livestock unit	0.099***	0.021	0.019	0.015	0.008	0.010	−0.005	0.008
Per capita cultivatable land	0.026	0.022	0.049*	0.030	0.003	0.010	0.022***	0.008
Exogenous income (I)	8.5×10^{-7}	1.4×10^{-6}	1.6×10^{-6}	2.1×10^{-6}	1.1×10^{-6}*	4.7×10^{-7}	5.5×10^{-7}	4.6×10^{-7}
Market characteristics (Ω_H)								
Distance from paved roads	−0.009	0.006	−0.014***	0.005	0.001	0.004	0.002	0.002
Price to wage ratio	3.165***	1.285	4.085***	1.167	1.526***	0.632	2.129***	0.482
Diffusion parameters (Ω_D)								
Extension	0.022	0.017	0.013	0.019	0.020**	0.009	−0.003	0.007
Exposure	0.171***	0.082	0.044	0.069	0.103**	0.051	0.054	0.039
Relative experience (τ)	1.084***	0.227	0.830***	0.154	−0.168	0.140	0.022	0.074
Social capital (Ω_S)								
Membership density	0.034	0.040	0.027	0.036	−0.068***	0.025	0.038**	0.017
Leader heterogeneity	0.168***	0.064	0.050	0.050	0.087**	0.040	0.076***	0.031
Norms of decisionmaking	0.220***	0.087	0.230***	0.068	0.014	0.061	−0.015	0.046
Net labor transfers	6.8×10^{-6}	5.2×10^{-6}	-3.0×10^{-6}	3.7×10^{-6}	5.8×10^{-7}	3.12×10^{-6}	1.2×10^{-6}	3.63×10^{-6}
Net cash transfers	2.3×10^{-6}*	1.23×10^{-6}	1.5×10^{-7}	9.3×10^{-7}	-6.2×10^{-7}	5.3×10^{-7}	1.6×10^{-8}	4.73×10^{-7}
Net others transfers	3.8×10^{-6}**	1.5×10^{-6}	1.5×10^{-6}	1.3×10^{-6}	-1.7×10^{-6}*	9.8×10^{-7}	-3.4×10^{-7}	6.2×10^{-7}
Mills ratio					0.001**	0.000	7.7×10^{-8}	5.5×10^{-8}
Constant					−0.086	0.461	−0.416	0.352
F (26, 104)					4.44		5.89	
Probability > F					0		0	
R^2					0.5258		0.4387	
Adjusted R^2					0.4073		0.3642	
Number of observations	332	=	332	=	131		223	

Table 7.3 (continued)

Variable	Probit model: Manure		Probit model: Mulch		Second step of Heckman model: Manure		Ordinary least squares model: Mulch	
	Marginal effects	Standard error	Marginal effects	Standard error	Coefficient	Standard error	Coefficient	Standard error
Observed probability	0.416		0.690					
Predicted probability at mean	0.399		0.785					
Correctly classified	79.2		78.1					
Likelihood ratio χ^2 (25)	124.24	=	120.18	=				
Probability > χ^2	0	=	0	=				
Log likelihood	–163.26		–145.518					
Pseudo R^2	0.2756	=	0.2922	=				

Note: ***, **, and * indicate statistically significant differences at the 1, 5, and 10 percent levels, respectively.

findings from other studies that slope is positively associated with perceptions about soil fertility problem, which in turn encourages use of soil fertility management practices (Ervin and Ervin 1982; Shiferaw and Holden 1998; Mwakubo et al. 2004).

One explanation for this finding relates to the properties of the technologies. Although erosion potential may encourage the use of erosion control technologies, such as conservation structures, it may act against the use of technologies whose benefits are likely to be lost when erosion potential is high. Such is the case with manure in banana production. On high slopes, the water runoff can easily wash manure out of the plantation, reducing its benefits. The capacity of the soil to retain moisture is a statistically significant factor only in the probability of using mulching practices. The negative sign can be interpreted in the same way as in the case of slope with manure. Benefits of incurring the high costs of mulching may be low when the capacity of the soil to retain moisture is low. Combined, the two results suggest a "Malthusian scenario" that people may perceive a soil fertility problem but do nothing about it. Poor drainage conditions of the soil affect only the probability of applying manure. Again, the positive association between poor drainage conditions and probability of using manure lends itself to the same explanation given for the findings regarding slope. Poor drainage conditions in the banana plots can also act through disease risk, because manure—and hence good plant nutrition—increases plant vigor that enhances tolerance to leaf spot diseases.

Most of the household characteristics (demographic factors, education, and wealth assets) were significant in the management decisions. As expected, education has a positive effect, reflecting the high degree of knowledge that these soil fertility management practices require. Other studies have also found a positive association between education and adoption of soil conservation technology (Ervin and Ervin 1982; Mwakubo et al. 2004). Endowment of wealth assets is equally important in decisions regarding use of soil fertility management practices. Ownership of livestock may reduce the cost of using manure, hence the positive effect of this parameter. Estimated effects of the per capita size of cultivatable landholding (adjusted for area in swamps and water bodies) have a positive sign but are only significant in the decision to use mulch, which reflects the fact that mulching is made on the farm using farmer resources. Although landholding size increases the availability of mulching materials, thus enabling farmers endowed with this physical asset to overcome the problem of the missing market, scale of production (represented by areas allocated to bananas) increases the incentive for mulching or manure by reducing fixed costs per unit areas (Feder and O'Mara 1981). Access to income from private assets does not seem to be important in the decision to use soil fertility management practices. This finding reinforces the point that these technologies are not introduced so much as they are made on-farm. Consequently, farmers with income from outside sources do not have any particular advantages in gaining access to the technology. However, results suggest that better access to bilateral transfers in the form of cash and gifts is positively associated with the probability of using soil fertility management practice (though it is only significant for manure use). This result could reflect the effect of social networks on risk aversion rather than liquidity constraint.

The age of the banana production decisionmaker and the dependency ratio in the household are negatively associated with the probability of manure application and use of mulching. Age is associated with a short time preference, which reduces the benefits of soil conservation technologies (Shiferaw and Holden 1998). A negative sign on the dependency ratio reflects labor constraints. Household size was positively associated with the probability that the household will choose to use mulch. In the presence of output and labor market imperfections, household size represents both the demand effects as well the effect of family labor endowment. Hence it is not clear which market imperfections explain the result.

The banana price relative to the unskilled labor wage rate in the village was significant in both technologies, perhaps because of the crucial role played by market incentives in crop management decisions. After controlling for the market prices, the probability of using mulching or manure practices seems to decrease with market access in general, reflected in the negative effect of distance from paved roads on the probability of use. The negative effect of distance from paved roads on the probability of using soil fertility management practices may be related to soil fertility depletion near the paved roads because of intensive banana commercialization. This intensification may increase the farmer perceptions of soil fertility problem, hence inducing a higher probability of using soil fertility management practices. Nkonya et al. (2004) found a similar result for the effects of market access on soil nutrient depletion in eastern Uganda.

Formal mechanisms of information diffusion (represented by exposure to new banana varieties and cumulative number of contacts with extension services) had the expected signs but were not statistically significant in most of the regressions, except for exposure in the use of manure. The low importance of formal information diffusion mechanisms in the probability of using soil fertility management in the Uganda rural communities is not surprising. Most of the households surveyed had never had contact with extension services regarding banana management. The cumulative number of contacts with extension services was on average only 1.88, a level too low to make any significant impact on decisions regarding use of knowledge-intensive technologies.

Instead, informal mechanisms of information diffusion were important determinants of the decision to use mulch or manure. Experience with the technology positively influences its increased use. Results also suggest that the number of household members that join associations has no effect on the probability that a household will use a soil fertility management practice. However, leadership heterogeneity in village associations is positively associated with the probability of use, but the variable was only significant in manure application. As argued by Rogers (1995), leaders act as opinion leaders, and when they are more educated and wealthier than others, they are likely to pull more information from outside the village because they are connected to better social networks. In Tanzania, Broeck (2004) found that farmers with secondary education and those with larger landholdings were more likely to seek information from outside their villages. Participatory norms of decisionmaking also show positive association with the probability of mulching and manure application, which is consistent with the findings of Isham (2000) in rural Tanzania that consultative norms positively influence adoption of fertilizers.

Extent of Use of Soil Fertility Management Practices

The second aspect of the use decision for a divisible technology is the extent of use. Extent of use was estimated using two methods: ordinary least squares regression and the Heckman model. Results from the estimation are presented in Table 7.3, with reasonable measures of fit for cross-sectional estimations. Most of the factors that were significant in the discrete decision to use management practices were also significant in deciding the extent of use. Some factors show contrasting effects in the two decisions, justifying the decision to estimate them separately. Results from the estimation of the extent of use are also technology specific.

Although elevation does not seem to influence the probability of using soil fertility management practices, the share of bananas grown under mulch is comparatively lower in high-elevation areas. One feasible explanation could be related to the farming system, which affects access to materials for mulch. Annual crops are more extensively grown in low-elevation areas, which could improve access to crop residues for mulching bananas groves in these locations relative to high-elevation areas. However, some experts consider that residues from annual crops generally provide only minimal material for mulching, with the occasional exception of maize stover.

Among the physical factors of land, only the slope of the farm appears to be important in the extent of soil fertility management. The variable was positively related to the extent of mulching with a larger magnitude (0.122). This result indicates that soil erosion potential that threatens loss of soil moisture will stimulate extensive use of technologies related to mulching. Scale of production influences both the probability and extent of mulching but in opposite directions. Households with larger banana groves apply soil fertility management practices to a smaller proportion of their bananas, perhaps reflecting the lower economic impact of soil depletion on farms with bigger banana groves. The finding is consistent with the prior information that agricultural growth in Uganda, as in many other Sub-Saharan African countries, has tended to rely on expansion in area rather than on productivity (MAAIF and MFEPD 2000).

In contrast, farmers with relatively large amounts of cultivatable land per capita apply mulch to a larger share of bananas than those with less land. The effect of land per capita on the extent of manure use is not statistically significant. Results suggest that an increase in population pressure will reduce the extent of mulching (with crop residues or grass) but has no apparent effect on use of manure. Thus, in the absence of appropriate technologies and policy incentives, a decrease in cultivable area per capita (a measure of population pressure on land) alone may not be sufficient to increase land-improvement investments in these areas, contrary to Boserup's (1965) hypothesis.

Household characteristics are also important determinants of the extent of soil fertility management for land allocated to banana production. Education of the production decisionmaker is positively related with the extent of mulching technology. As in the case of the discrete decision, household size positively influences use of soil fertility management practices. As already pointed out, in the presence of a combination of labor and output market imperfections, it is not clear which market imperfections are implied by the results. Older farmers are likely to allocate a smaller share of their banana groves to manure because of their time preferences (Shiferaw and Holden 1998). Other household demographic characteristics (such as gender and dependency ratio), as well as livestock wealth, do not appear important in the extent of soil fertility management. Similarly, exogenous income was only significant in manure application. This finding could be feasibly explained by the fact that manure, unlike mulching, requires expensive implements, such as wheelbarrows and spades, to implement the technology, which may discourage a household with liquidity constraints from extensively using the technology.

As expected, the estimates on the variable price-wage ratio indicate that the greater the return to banana production relative to the opportunity cost of labor, the higher the use of fertility-enhancing practices, which require labor. This is a key finding. Improved banana management practices are labor intensive, and higher costs of hired labor relative to the banana market prices will have a negative effect on the intensity of management. Thus rural development trends, such as urbanization, that increase the opportunity cost of labor are likely to have negative consequences on soil fertility management. After controlling for market prices, the distance from the paved roads and indicators of market access in general are not significant at conventional levels in both technologies, suggesting that general increase in market access has no effect on the extent of banana management using mulch or manure.

Results also indicate that social capital is an important determinant of the extent of soil management in banana production. Household density of membership in associations was associated with the extent of mulching and manure application but with opposite signs. The finding that household membership density in associations influences the extent of mulching but does not influence the probability of use is interesting. The result suggests that this aspect of social capital operates through various complementary mechanisms to influence use of this technology. One possible interpretation is that, in addition to information, membership density may work through peer pressure. Mwakubo et al. (2004) also found that in the marginal areas of Kenya, membership density increases intensity of soil conservation technologies. On the contrary, estimation results indicate that increase in the number of household members joining associations is associated with reduced demand for manure. Perhaps this result implies that more household members joining associations might reduce labor available for crop production, generating consequences that might exceed the positive benefits for more labor-intensive techniques. Manure application by Ugandan banana producers involves gathering and transporting heavy animal manure from the kraal to plots or composting kitchen residues in pits. This effort may require additional labor compared to mulching with crop residues.

Households in villages where norms of decisionmaking in associations are participatory and where leaders are wealthier or better educated than most group members are likely to apply management practices to a larger share of mats in their banana groves. This finding is consistent with the work of Rogers (1995) and later Isham (2000) that village leaders with capacity to network enhance the adoption decisionmaking, because they act as opinion leaders. Better-connected leaders may mobilize more extensive links to new information on farming methods (Granovetter 1973), thereby reducing the time lag to adoption.

The role of bilateral transfers from social networks in the extent of mulching and manure use decisions was also assessed. Net transfer of labor, cash, and other consumer durable goods were included as measures of resources accessed from social networks. Most forms of the bilateral transfers from social networks were not significant in decisions on the extent of use, except net transfers in the form of gifts, which were negatively associated with manure application. One possible interpretation for the negative sign is that people who receive gifts are poor, and they use the transfers on their basic needs rather than on banana production. Hogset (2005) also found a weak, though positive, association between bilateral transfers and the adoption of soil conservation technologies in Kenya. Lack of statistical significance for most of the forms of bilateral transfers is an indication that social capital does not perfectly substitute for markets in agriculture development.

Conclusion

The first conclusion from this analysis is that soil fertility management decisions of banana producers depend on a host of factors. Second, the analysis demonstrates the significance of a range of factors related to social capital in soil fertility management decisions, confirming the importance of investigating the role of such factors in agricultural development. Results provide evidence that the density of membership in associations, the characteristics of those associations, and the amount of bilateral transfers accruing to the household from its social networks significantly influence soil fertility management decisions. The nature of the relationship of these factors to the management practices depends on the type of the technology and also on the specific form of social capital used in the analysis. This finding supports the conclusion that accounting for social capital in development projects should be based on a thorough analysis of the institutional context, relating this context to the properties of the technology.

The results also indicate that density of household membership positively influences the extent of mulching, implying that where the promotion of mulching is the main objective, household members should be encouraged to participate more in associations while minimizing the negative consequences on the use of other production techniques. Through interactions in associations, members may share information from their experiments. Because the information gained comes from other farmers whose opinion the adopter trusts, the potential adopter may skip the stage of experimentation and hence adopt the technology extensively. In addition, when more members participate in associations, they can pool information, so that the influence of the association carries over to the farm, persuading those who manage banana plants to conform.

Characteristics of associations are the other aspect of informal institutional social networks that were hypothesized to be important in soil management decisions. Two characteristics were examined: (1) the capacity of leadership, represented by heterogeneity of leadership in terms of education and livelihood; and (2) norms of decisionmaking. Results support the prior expectation that capacity of leadership and norms of decisionmaking are important in soil management decisions. The most important policy conclusion from this result is that interventions that use group-based approaches should consider sensitizing people to the role and importance of leaders and participatory decisionmaking. This conclusion has implication for the extension methodology. Extension sevices should not only emphasize the benefits of the technology but also include messages that prepare the community for meaningful and effective interactions.

Although bilateral transfers are statistically significant in decisions about soil fertility management practices, the magnitude of coefficients is too small to matter much. The most important conclusion from this result is that although private social networks may be used to compensate for costs of market transactions, these do not overcome market constraints with respect to soil conservation investments.

In addition to social capital, the research has identified other factors that need the attention of policymakers. These include market infrastructure, educational programs that influence farmers' perceptions of the soil fertility problem, and knowledge about management practices and poverty in general. Improving access to markets for banana producers, raising the output price relative to the cost of labor, is likely to have the greatest impact of any single factor. Results provide strong evidence that market incentives increase both the probability of use and extent of use. A major implication of these findings is that banana growers in Uganda respond to market incentives.

Household endowments of family labor and other production assets (such as livestock and per capita land availability) are critical for good management of bananas, reflecting market imperfections for these factors. Removing market imperfections completely is a long-term undertaking. In the meantime, there is need to identify technologies that demand less of farmers' resources. Such technologies should also address the problem of increasing population pressure on the land. For example, because there is no market for mulching materials, the extensive use of this practice depends on the availability of land. Hence, use of this practice is negatively related to increased population pressure.

The issue of soil fertility depletion in areas with greater access to road infrastructure also needs more attention for long-term banana productivity improvement. Although general improvement in access to markets increases the probability of using soil fertility management practices, it does not seem to motivate the extent of the use, due to higher opportunity costs of labor. This finding is of importance, given that improving banana productivity through soil fertility management will not be achieved based on high rates of use only but on the extent of their use. Moreover, the problem of soil fertility depletion is likely to be intensive in these areas, given the high levels of banana commercialization.

High erosion potential encourages use of mulching practices but discourages use of manure, whose production function shifts inward when the erosion potential is perceived to be high. This result suggests that other techniques that reduce erosion potential need to be promoted along with manure, while the positive role of mulching on curbing erosion potential should be emphasized in extension messages. Finally, results support intensification of educational programs as a means of promoting the use of soil fertility management practices. Hence, more support to extension programs would increase use of these techniques.

In the context of the research presented in this report, it should be emphasized that long-term improvement in banana production in Uganda will be achieved by relieving not only the constraints related to both pests and disease (addressed in Chapter 6), but those related to soil fertility. This chapter is an initial attempt to better understand the factors that influence whether banana farmers invest in soil fertility management. Chapter 9 directly examines the relationships of soil fertility and labor to banana productivity.

References

Barr, A. 2000. Social capital and technical information flows in the Ghanian manufacturing sector. *Oxford Papers* 52 (3): 539–559.

Boserup, E. 1965. The conditions of agricultural growth. In *The economics of agrarian change under population pressure.* New York: Aldine.

Broeck, V. 2004. Social effects in growing bananas: Evidence from a Tanzanian village. Katholieke Universiteit Leuven, Leuven, Belgium. Manuscript.

Ervin, C. A., and E. D. Ervin. 1982. Factors affecting the use of soil conservation practices: Hypothesis, evidence and policy implications. *Land Economics* 58 (3): 277–292.

Fafchamps, M., and S. Lund. 2003. Risk-sharing networks in rural Philippines. *Journal of Development Economics* 71 (2): 261–287.

Feder, G., and G. T. O'Mara. 1981. Farm size and the diffusion of green revolution technology. *Economic Development and Cultural Change* 30 (1): 59–76.

Feder, G., and R. Slade. 1984. The acquisition of information and the adoption of new technology. *American Journal of Agricultural Economics* 66 (3): 312–320.

Feder, G., and D. Umali. 1993. The adoption of agricultural innovations: A review. *Technology Forecasting and Social Change* 43: 215–239.

Feder, G., R. E. Just, and D. Zilberman. 1985. Adoption of agricultural innovations in developing countries: A survey. *Economic Development and Cultural Change* 33 (2): 255–298.

Granovetter, M. S. 1973. The strength of weak ties. *American Journal of Sociology* 78 (6): 1360–1380.

Greene, W. H. 2000. *Econometric analysis.* Upper Saddle River, N.J.: Prentice-Hall.

Hogset, H. 2005. Social networks and technology adoption. Selected paper, American Agricultural Economics Association Annual Meeting, July 1–3, Providence, R.I., U.S.A.

Isham, J. 2000. A model of technology adoption with social capital. Ph.D. dissertation, University of Maryland, College Park, U.S.A.

Kennedy, P. 1985. *A guide to econometrics.* Cambridge, Mass.: MIT Press.

MAAIF and MFEPD (Ministry of Agriculture, Animal Industry and Fisheries and Ministry of Finance, Economic Planning and Development). 2000. *Plan for the modernization of agriculture: Eradicating poverty in Uganda.* Kampala, Uganda: MAAIF and MFEPD.

Maddala, G. S. 1983. *Limited dependent and qualitative variables in econometrics.* London: Cambridge University Press.

Mwakubo, S. M., G. A. Obare, J. Omiti, and L. Mohammed. 2004. The influence of social capital on sustainable agriculture in marginal areas in Kenya. Eastern Africa Food Policy Network Report 13. Kampala, Uganda: International Food Policy Research Institute.

Nkonya, E., J. Pender, P. Jagger, D. Sserunkuma, C. Kaizzi, and H. Ssali. 2004. *Strategies for sustainable land management and poverty reduction in Uganda.* Research Report 133. Washington, D.C.: International Food Policy Research Institute.

Pender, J., and J. Kerr, 1996. Determinants of farmer's indigenous soil and water conservation investments in India's semi-arid tropics. Environment and Production Technology Division Discussion Paper 17. Washington, D.C.: International Food Policy Research Institute.

Rogers, E. M. 1995. *Diffusion of innovations.* New York: Free Press.

Shiferaw, B., and S. Holden. 1998. Resource degradation and adoption of land conservation technologies in the Ethiopian Highlands: A case study in Andit Tid, North Shewa. *Agricultural Economics* 18 (3): 233–247.

Stevens, J. P. 2002. Applied multivariate statistics for the social sciences. Fourth edition. Mahwah, N.J.: Lawrence Erlbaum Associates.

Tushemereirwe, W. K., I. N. Kashaija, W. Tinzaara, C. Nankinga, and S. New, eds. 2003. A *guide to successful banana production in Uganda. A banana production manual.* Kampala, Uganda: National Banana Research Programme, Kawanda Agricultural Research Institute.

Wooldridge, W. J. 2002. *Econometric analysis of cross section and panel data.* Cambridge, Mass.: MIT Press.

Yli Renko, H., E. Autio, and V. Tontti. 2002. Social capital, knowledge, and the international growth of technology-based new firms. *International Business Review* 11 (3): 279–304.

CHAPTER 8

Determinants of Banana Productivity and Technical Efficiency in Uganda

Fredrick Bagamba, Ruerd Ruben, and Mariana Rufino

This chapter analyzes the technical efficiency of banana production among Ugandan smallholders by estimating a stochastic production frontier model with inefficiency effects. The empirical models are formulated within the overarching framework of the agricultural household specified in Chapter 2. The data design is also summarized in Chapter 2, although additional information about soil quality was collected for this analysis from three sites selected for this purpose. Hypotheses are tested about the regional effects on productivity relationships and returns to scale. Specific hypotheses are tested with respect to the impact of market access and household- and farm-specific factors on technical efficiency. This chapter examines banana productivity as it relates to two important constraints in banana production: soil fertility and labor; it therefore complements the analysis in Chapter 6, which treats biotic constraints and the potential adoption of resistant cultivars, and Chapter 7, which examines the adoption of soil fertility management practices by farmers.

Stochastic Frontier Production Function

A large body of theoretical and empirical literature has investigated the measurement of efficiency of farm enterprises, using various methods. Ali and Byerlee (1991) have emphasized that the focus in analyzing economic efficiency should be the performance of the whole production system, including farmers and institutional support systems. These results can be used to pinpoint the factors that impede the capacity of farmers to reach their productivity potential.

Technical efficiency (TE) can be estimated using one- or two-step approaches. In the two-step procedure, the production frontier is estimated first, and the technical efficiency of each firm is derived subsequently. In the second step, the derived technical efficiency variable is regressed against a set of variables that are hypothesized to influence the firm's efficiency (Kalirajan 1981; Pitt and Lee 1981). However, the two-stage procedure lacks consistency in assumptions about the distribution of the inefficiencies. In step one, it is assumed that inefficiencies are independently and identically distributed in order to estimate their values. In step two, estimated inefficiencies are assumed to be a function of a number of firm-specific factors, violating the assumption in step one (Coelli, Rao, and Battese 1998). To overcome this inconsistency, Kumbhakar, Gosh, and McGuckin (1991) suggest estimating all the parameters in one step. In a one-step procedure, which we adopt for this study, the inefficiency effects are defined as a function of the farm-specific factors and incorporated directly into the maximum likelihood (ML) estimate.

TE is measured as a ratio of actual to potential output (Aigner, Lovell, and Schmidt 1977; Meeusen and van den Broeck 1977). Approaches for measuring TE generally vary from programming (nonparametric) approaches to statistical estimation (parametric) approaches, depending on functional forms and techniques for estimating the potential output (Forsund, Lovell, and Schmidt 1980; Bauer 1990; Fried, Lovell, and Schmidt 1993; Coelli 1995; Kalirajan and Shand 1997). In analyzing farm-level data where measurement errors are substantial and weather is likely to have a significant effect, the stochastic frontier method is usually recommended (Coelli 1995).

Early frontier production functions that followed Farrell (1957) were deterministic in that they assumed a strict one-sided error term (Schmidt 1986; Coelli 1995). One of the major criticisms against deterministic frontier estimates is that no account is taken of the possible influence of the measurement errors and other data noises on the shape and the positioning of the estimated frontiers. All the observed deviations from the estimated frontier are assumed to be a result of technical inefficiency (TI; Coelli, 1995). Aigner, Lovell, and Schmidt (1977) and Meeusen and van den Broeck (1977) proposed a stochastic frontier production function, in which sources of data noise are accounted for by adding a symmetric error term to the non-negative error. The parameters of this model are estimated by ML, given suitable distributional assumptions for the error terms. The stochastic frontier is not, however, without problems. The major limitation is that one has to make arbitrary assumptions regarding the functional form of the frontier and the distributional form of the error. Moreover, as the model is estimated by ML, the solution obtained might not be optimal, because the likelihood function is not globally concave and allows for multiple local maxima (Maddala 1971).

Using the statistical estimation approach, we define a farm specific stochastic production frontier involving outputs and inputs as:

$$y_i^* = f(x_i)\exp(v_i) \tag{1}$$

where y_i^* is the maximum possible stochastic potential output from the ith farm, x_i is a vector of m inputs, and v_i are statistical random errors assumed to be distributed as $N(0,\sigma_v^2)$. The production realized on the ith farm can be modeled as:

$$y_i = y_i^* \exp(-u_i) \tag{2}$$

where y_i^* is the maximum possible stochastic potential output from the ith farm, x_i is a vector of m inputs, and v_i are statistical random errors assumed to be distributed as $N(0, \sigma_v^2)$. The production realized on the ith farm can be modeled as:

$$TE = \exp(-u_i) = \frac{y_i}{y_i^*}. \tag{3}$$

Substituting equation (1) into equation (2) and taking logs on both sides, we get:

$$\ln y_i = \ln f(x_i) + v_i - u_i, \tag{4}$$

where y_i denotes the production of the ith farm ($i = 1,2, \ldots, n$); x_i is a $(1 \times k)$ vector of functions of input quantities used by the ith farm; each is assumed to be an independently and identically distributed random error independent of every u_i, and u_i is a one-sided error term representing the technical inefficiency of farm i.

Subtracting v_i from both sides of equation (4), the production of the ith farm can be estimated as:

$$\ln y'_i = \ln f(x_i) - u_i. \tag{5}$$

The efficient level of production can be defined as:

$$\ln \hat{y}_y = \ln f(x_i)\,. \qquad (6)$$

From equations (5) and (6), we can compute TE given by:

$$\ln TE_i = \ln y' - \ln \hat{y} = -u_i \qquad (7)$$

$TE_i = e^{-u_i}$ and is constrained to be between 0 and 1. When $u_i = 0$, then $TE = 1$ and production is said to be technically efficient.

The distribution of u_i could be half normal with zero mean, truncated normal (at mean μ), or based on conditional expectation of the exponential $(-u_i)$. There are no a priori reasons for choosing a specific distributional form, because each has advantages and disadvantages (Coelli, Rao, and Battese 1998). The half normal and exponential distributions have a mode of zero, implying that most firms being analyzed are efficient. The truncated normal allows for a wide range of distributional shapes, including nonzero modes, but is computationally more complex (Coelli, Rao, and Battese 1998).

We adapt the model proposed by Battese and Coelli (1995), in which the TI effects are defined by:

$$u_i = \mathbf{z}_i\boldsymbol{\delta} + w_i\,, \qquad (8)$$

where $\mathbf{z}_i$ is a $(1 \times m)$ vector of explanatory variables associated with the TI effects; $\boldsymbol{\delta}$ is a $(m \times 1)$ vector of unknown parameters to be estimated; and w_i is an unobservable random variable. The parameters indicate the impacts of variables in $\mathbf{z}$ on TE. A negative value suggests a positive influence on TE and vice versa. The frontier model may include intercept parameters in both the frontier and the model for the inefficiency effects, provided the inefficiency effects are stochastic and not merely a deterministic function of relevant explanatory variables (Battese and Coelli 1995).

The null hypothesis that the TI effects are not random is expressed by H_0: $\sigma_v = 0$. Accepting the null hypothesis that $\sigma_v = 0$ would indicate that σ_u^2 is zero and thus the term u_i should be removed from the model, leaving the specification that can be consistently estimated by ordinary least squares (Coelli 1994). Further, the null hypothesis that the impact of the variables included in the inefficiency effects model in equation (8) on the TI effects is zero is expressed by H_0: $\boldsymbol{\delta}' = \mathbf{0}$, where $\boldsymbol{\delta}'$ denotes the vector ($\boldsymbol{\delta}$) with the constant term (δ_0) omitted, given that it is included in the expression $\mathbf{z}_i\boldsymbol{\delta}$ (Battese and Broca 1997).

Factors Affecting Technical Efficiency

In crop production, TE is likely to be affected by a wide range of farm- and village-specific factors. Forsund, Lovell, and Schmidt (1980) argue that inefficiency is typically related to factors that are associated with farm management practices. Such factors include education, family size and composition, experience, proximity to markets, and access to credit. Education, which is directly related to management skills, has received adequate attention in the efficiency literature (Weir 1999; Tian and Wan 2000; Weir and Knight 2000; Binam et al. 2003). The results of the impact of education on TE are mixed, with some studies showing positive impact (Belbase and Grabowski 1985; Kalirajan and Shand 1986; Bravo-Ureta and Pinheiro 1997) and others showing a negative impact (Kalirajan 1984, 1991; Phillips and Marble 1986; Bravo-Ureta and Evenson 1994). Education increases the household's ability to utilize existing technologies and attain higher efficiency levels (Battese and Coelli 1995). In our study, we use education of household as a proxy for management skills and age of household head as a proxy for experience (learning by doing). TE is expected to increase with age as the farmer gains experience, but at a decreasing rate as the farmer becomes elderly. Access to resources (land, labor, and capital) is one of the reasons for this type of behavior. Young households are deficient in resources and might not be able to apply inputs or implement certain agronomic practices sufficiently quickly. Timely application of inputs and implementation of management is expected to enhance efficiency. The other factor that explains the quadratic relationship between age and efficiency reflects access to information. Elderly farmers are less likely to have contacts with extension and training programs, and are therefore less willing to adopt new practices and modern inputs (Hussain 1989).

Gender of the household head is expected to have significant effects on TE. Farms managed by men are expected to attain higher TE than those managed by women. Men have more access to resources and information and are more likely to obtain credit, which increases production efficiency.

The effect of household size on TE has not been widely reported in the literature. Household size is expected to influence TE through its effect on the labor endowments of households (including child labor). Large households are expected to be more technically efficient, because they can implement activities on time, attaining higher output with the same or less labor input. The effect of more adults per household on TE is expected to produce mixed results. On the one hand, an increase in the number of adults in the family could increase TE if it results in increased labor devoted to banana production. On the other hand, the effect could be negative if adults have higher chances of obtaining off-farm employment. The effect could be insignificant if labor withdrawn from the farm into off-farm employment is substituted with capital inputs.

Another factor for which the effect on TE has been infrequently reported in the literature is proximity to factor markets. Households located nearer factor markets are expected to have higher TE than those located in remote areas. Proximity to good roads increases access to training and extension programs, from which farmers can attain information and skills for better crop management. Proximity to markets also increases farmers' access to credit facilities and income-generating activities (such as off-farm employment) that enable them to buy and apply inputs on time. By contrast, access to markets may increase the access farmers have to alternative employment with higher returns than from farming, leading them to reallocate labor from farm to nonfarm activities. Households located in remote areas that have greater access to farm labor are expected to attain higher efficiencies than do households in close proximity to nonfarm labor markets.

Production Function

A number of functional forms have been used in the empirical estimation of frontier models. The simplest, the Cobb-Douglas, is specified in logarithmic form as:

$$\ln y = \ln A + b_1 \ln x_1 + b_2 \ln x_2 . \quad (9)$$

The transcendental production function, which is a generalized Cobb-Douglas function, is:

$$\ln y = \ln A + b_1 \ln x_1 + b_2 \ln x_2 + c\, {}^{x_1}\!/_{x_2} . \quad (10)$$

A more complex form, the transcendental logarithmic (translog) form is:

$$\ln y = \ln A + \sum_i b_i \ln x_i + \tfrac{1}{2}\sum_i \sum_j b_{ij} \ln x_i \ln x_j . \quad (11)$$

The most commonly used function forms are the translog and Cobb-Douglas function. Often preferred for its simplicity, the Cobb-Douglas imposes restrictions on returns to scale and elasticities; the translog imposes no restrictions on returns to scale but suffers from multicollinearity and degrees-of-freedom problems. In any case, the impact of functional form on estimated efficiency has been reported to be very limited (Kopp and Smith 1980). Battese and Broca (1997) recommend approaches in which more general model specifications and assumptions are made and simpler formulations are formally tested. In our estimations of the frontier production functions, we use each of the three function forms to estimate the production of cooking bananas. We then compare the results of the inefficiency effects across the three forms.

Accounting for Soil Nutrients and Organic Matter

Agricultural production in Uganda, as in many other developing agricultural economies, depends largely on land and labor input, with little or no external inputs used. The soils are poor in nutrients and rely on recycling of nutrients from soil organic matter (SOM) to maintain crop productivity. The soil's ability to retain and supply nutrients to a crop depends on the cation exchange capacity (CEC)—soils with high CEC are able to bind more cations, such as K^+, to the exchange sites of clay and SOM particle surfaces. Soils with high CEC also have a greater battering capacity and thus the ability to resist changes in pH. Thus soils with high amounts of clay and/or SOM typically have higher CEC and buffering capacities than do more silty or sandy soils. Soil pH also affects nutrient retention and availability to crops. Soils with high pH have low concentrations of H^+, which enables more base cations to be on the particle exchange sites, thus making the soil less susceptible to leaching. With the exception of P, which is most available within a pH range of 6 to 7, other macronutrients (N, K, Ca, Mg, and S) are more available within a pH range of 6.5 to 8. High rainfall can result to soil acidity (Tisdale et al., 1993). Rufino (2003) found that unfavorable soil pH limits maximum yield in 42 percent of the banana plots in Bamunanika, Kisekka, and Ntungamo, which is indicative of other soil fertility problems. In the same sites, soil K was a limiting factor for 19 percent of the banana plots, N was limiting in 12 percent, whereas P was not a limiting factor. Exchangeable K is determined by the neutral ammonium acetate method (Thomas 1982). Available P is determined by the Olsen method (Olsen et al. 1954).

There is a need to take into considerations the interrelations between N, K, SOM, soil texture, and chemical characteristics in modeling production behavior. First, SOM is affected by the soil texture and drainage (sand content), C:N ratios of organic materials, climate, and cropping practices. The SOM content can be estimated as:

$$\ln SOM = \alpha_0 + \alpha_1 sand + \alpha_2 D1 + \alpha_3 D2, \quad (12)$$

where ln *SOM* is the natural log of soil organic matter content (percent), $\alpha_0, \alpha_1, \alpha_2, \alpha_3$ are parameters to be estimated, *sand* is the ratio of sand to (sand + clay + silt; percent), and *D*1 and *D*2 are village dummies for measuring the impact of differences in climate and cropping practices. Equation (12) can be estimated by ordinary least squares to obtain the estimates of $\alpha_0, \alpha_1, \alpha_2, \alpha_3$.

Soil N is highly correlated to SOM, organic amendment (mainly animal manure), and regional characteristics and can be estimated as:

$$\ln N = \theta_0 + \theta_1 \ln SOM + \theta_2 M + \theta_3 D1 + \theta_4 D2, \quad (13)$$

where ln *N* is the natural log of soil N content (percent), $\theta_0, \theta_1, \theta_2, \theta_3, \theta_4$ are parameters to be estimated, and *M* is animal manure input (kg/year). The remainder of the variables are as already defined. Equation (13) can be estimated using a two-stage least squares model in which ln *SOM* is instrumented by *sand*.

Availability of soil K is affected by soil pH, SOM content, and additions of crop residues and can be estimated as:

$$\ln K = \delta_0 + \delta_1 pH + \delta_2 \ln SOM + \delta_3 C + \delta_4 D1 + \delta_5 D2, \quad (14)$$

where ln *K* is the natural log of available soil K (meq/100 g soil), δ_{0-5} are parameters to be estimated, *pH* is soil pH, and *C* is crop residue input (kg/year). Equation (14) is estimated using two-stage least squares, again instrumenting ln *SOM* with the *sand* variable.

Crop output is determined by labor input, area allocated to the crop, and nutrient availability (mainly N and K for bananas). Organic amendment (animal manure, grass mulch, and crop residues) contribute to soil nutrients but also to the physical and chemical properties of soil, enabling a given land area to produce higher output. Crop output can be modeled as:

$$\ln Y = \beta_0 + \beta_1 \ln A + \beta_2 \ln L + \beta_3 M \cdot$$
$$\beta_4 C + \beta_5 \ln SOM + \beta_6 \ln K \cdot$$
$$\beta_7 D1 + \beta_8 D2 \qquad (15)$$

where ln Y is the natural log of crop output (kg/year), β_{0-8} are parameters to be estimated, ln A is the natural log of the area allocated to crop (cooking bananas; in acres), and ln L is the natural log of labor input (hours/year). Equation (15) can be estimated using two-stage least squares, with *sand* as the instrument for *SOM* and *pH* as that for for *K*.

To obtain efficient estimates, equations (12), (14), and (15) are estimated simultaneously using a three-stage least squares method, which is the most appropriate technique to use to estimate a system of equations with endogenous variables included on the right hand side.

The three equations (12), (14), and (15) can be collapsed into a reduced-form equation:

$$\ln Y = \beta_0 + \beta_1 \ln A + \beta_2 \ln L + \beta_3 M$$
$$\beta_4 C + \beta_5 sand + \beta_6 pH$$
$$\beta_7 D1 + \beta_8 D2. \qquad (16)$$

Endogeneity

Equation (16) is estimated using ordinary least squares. A problem could arise if labor input were endogenously determined. We test for endogeneity by first estimating the labor equation with wage rate, output price, household characteristics, and opportunities included on the right hand side. The residual obtained from the estimated labor equation is then included on the right hand side in the production function estimation. If the effect of the residual turns out to be significant (5 percent), then labor input is confirmed as endogenously determined. An instrumental variable approach or two-stage least squares would be the approach to use to obtain efficient and consistent estimates if valid instruments are available. If the soil quality variables are included in equation (9), ordinary least squares is valid for obtaining consistent and efficient estimates of manure and other organic amendments. When soil quality variables (*sand* and *pH*) are missing in equation (16), the manure and crop residue variables can be treated as endogenous, because farmers would tend to apply these inputs where soils are poor, and no application is carried out where the soil is fertile. We lack sufficient and valid instruments for manure and crop residues. Therefore the estimates for manure and crop residue should be interpreted with care. In the absence of endogenous variables on the right hand side, equation (16) can be consistently estimated using a stochastic frontier approach.

Data

The data design is described in Chapter 2 and Appendix D. Of the total sample of 660 farmers surveyed in Uganda, data for 508 were usable in the analysis. The production function is estimated for cooking bananas, whereas the whole sample was selected for farmers that grow bananas. Some farmers, especially in the lower elevation areas, had banana plots that were less than 2 years old and harvested no output. Others had abandoned plots and did not allocate labor to them. These farms were not included in the estimation. Some households had missing cases in some of the variables, and therefore were excluded from the sample.

The final sampling frame consisted of 27 subcounties, of which 3 were selected (Ntungamo, Bamunanika, and Kisekka) to complement soil analyses. Ntungamo subcounty represents the high-elevation region, where banana production levels are high and there are no evident signs of decline, whereas Kisekka and Bamunanika sub-counties represent the low-elevation region, where there is serious decline in yield and production. The sample stratification enables us to capture elevation-related effects (differences in rainfall, temperature, pests, and disease pressure) on banana production (Chapter 2 and Appendix D).

Definitions of variables and summary statistics are shown in Table 8.1. The average area under cooking bananas is 0.58 acres. More area is allocated to cooking bananas in the high-elevation compared with that of the low-elevation region. The trend for allocation of labor and other inputs (manure, grass mulch, and crop residues) is the same, with farmers at high elevations applying greater quantities of inputs compared to those applied by farmers at low elevations.

Farm sizes are considerably larger in the low-elevation region compared to the highlands. Banana plantations are longer lasting at high elevation than at low elevation. Black Sigatoka and weevil scores are somewhat higher for low elevations compared to those for the high-elevation region. Heads of households are slightly more aged and more educated, and more households are located in remote areas in the low-elevation than in the high-elevation region. The average distance from the highway to farms is 14.6 km for low elevation and 10.3 km for high elevations. Farmers in the low-elevation region have higher access to credit compared to those in the highlands. The average amount received in remittances and rent is lower for the low-elevation region compared to the amount received in the high-elevation region.

Results

The hypothesis that labor is endogenously determined in the production of cooking bananas is rejected in the case of the high-elevation region but not for low elevations. The residual variable, included in the second stage of the two-stage least squares method, is found to have no significant effect in the case of high elevation, leading to the rejection of the hypotheses that labor is endogenous (Table 8A.1 in the Supplementary Tables section of this chapter). The endogeneity hypothesis assumes a two-way causal relationship in which farmers are thought to rely on the expected output in deciding about the amount of labor to allocate to production of cooking bananas. At the same time, the amount of labor allocated would determine the output obtained from the production process. Rejection of the endogeneity hypothesis implies that labor used in the production of cooking bananas is exogenously determined, independently of the expected output. However, for low elevations, labor input is most likely not predetermined, and estimates from the frontier production function need to be interpreted cautiously.

Results of the frontier function are shown in Table 8.2. Results from the Cobb-Douglas function show that output responds positively to area and labor in both regions, consistent with expectations. The results show that labor contributes more to productivity compared with crop area. The labor/crop area (L/A) variable has significant effect in the transcendental function for the low-elevation region but not for high elevations. Manure has a positive and significant effect on productivity. The effects of grass mulch and crop residues are only significant in the high-elevation region, where the effects are positive and significant. Farm size has a positive influence on productivity but the effect is only significant for the high-elevation region. For a given size of banana plot, farmers with larger farms have higher banana yields. Large farmers are more likely to be committed to farming than are small farmers, who are more likely to diversify into off-farm wage employment. Extension visits have a positive effect and are significant (1 percent) in the high-elevation but not in the low-elevation region. Interaction with

Table 8.1 Variable definitions and summary statistics for cooking bananas, productivity and technical efficiency analysis

		Overall sample		Low elevation		High elevation		Case study	
		n = 512		n = 374		n = 138		n = 157	
Variable	Definition	Mean	Standard deviation	Mean	Standard deviation	Mean	Standard deviation	Mean	Standard deviation
Y	Cooking bananas output (kg/year)	3,109.7	4,919	1,581	1,958	7,252	7,494	4,125.8	5,870.9
Variables in the production function									
A	Area (acres) under cooking bananas	0.58	0.76	0.504	0.771	0.801	0.688	0.622	0.569
L	Labor input (hours/year)	636.3	649.7	470	509	1,088	769	521.9	5,208
M	Manure input (kg/year)	495	2707	292	1245	1,045	4,765	528.2	2,610.3
G	Grass mulch input (kg/year)	194	1,461	95	672	461	2,577		
C	Crop residue input (kg/year)	331	1,660	280	1,114	469	2,620	205.2	1,008.5
N	Soil nitrogen (percent)							0.127	0.07
K	Available soil potassium (meq/100 g soil)							1.165	1.124
SOM	Soil organic matter (percent)							4.839	1.806
pH	Soil pH							5.975	0.619
sand	Ratio of sand to (sand + clay + silt) (percent)							59.35	9.99
farm size	household farm size (acres)	4.023	8.567	4.45	9.39	2.866	5.635		
Ext	Extension visits in 6 months	0.702	1.912	0.69	2.05	0.73	1.487		
plotage	Age of banana plot (years)	20	23	11.9	13.73	41.8	28.3		
$plotage^2$	Plotage squared	926	1,996	329.5	900.7	2,544	3,006		
Sigatoka	Sigatoka score[a]	0.163	0.272	0.22	0.3	0.02	0.07		
weevils	Weevil score[a]	0.394	0.333	0.42	0.34	0.33	0.32		
Technical efficiency variables									
Age	Age of household head (years)	45.2	16	45.72	16.46	43.7	14.6		
Age^2	Age squared	2,295	1,610	2,360	1,673	2,118.3	1,417.4		
edhh	Education household head (years)	5.39	4.09	5.55	4.12	4.93	3.99	5.981	4.1
D	Distance to tarmac road (km)	13.46	18.7	14.6	20.3	10.3	12.97		
hhsz	Household size	5.89	2.65	5.84	2.70	6.02	2.52	5.329	2.323
depr	Persons >64 or <14 years old/family size	0.497	0.239	0.498	0.252	0.494	0.201		
hplot	Plot managed by husband	0.764	0.425	0.73	0.445	0.854	0.354	0.839	0.369
kk	Amount credit obtained (thousand Ush)	14	92.3	17.83	107.3	3.66	17.15		
sk	Remittances + rent (thousand Ush)	90	368	80.65	306	115.3	500.1		
wp	Real wage rate (wage/price bananas)	2.67	1.08	2.72	1.21	2.53	0.58		

[a]Farmers were asked to score the presence of the disease or pest on a particular plot and the number of years the disease or pest had been observed on the plot. Presence was scored as 1 and absence as 0. The final score of the disease or pest was computed taking into consideration the number of years it had been observed on the plot and the size of the plot. For example, if the household has three plots with disease scores 0 for all the years, 1 for 3 years out of 5 years, and 1 for 7 out of 10 years and the corresponding areas of each plot are 0.5, 0.9, and 1.5 acres, the final score is (0 × 0.5 + 0.6 × 0.9 + 0.7 × 1.5)/(0.5 + 0.9 + 1.5) = 0.548.

Table 8.2 Results of the frontier function

	Overall sample			Low elevation			High elevation		
Variable	Eql	Eq2	Eq3	Eql	Eq2	Eq3	Eql	Eq2	Eq3
N	512	512	512	374	374	374	138	138	138
Constant	5.79***	5.158***	4.262***	5.694***	5.064***	4.279***	6.637***	6.259***	3.526
	(18.99)	(13.73)	(5.27)	(15.75)	(11.32)	(4.45)	(14.52)	(9.68)	(0.78)
$\ln(A)$	0.332***	0.207***	−0.211	0.332***	0.206***	−0.233	0.264***	0.199**	0.065
	(8.04)	(3.44)	(−0.77)	(6.45)	(2.81)	(−0.68)	(5.79)	(2.19)	(0.10)
$\ln(L)$	0.384***	0.489***	0.864***	0.395***	0.5***	0.812***	0.282***	0.343***	1.189
	(8.87)	(8.64)	(3.49)	(7.87)	(7.55)	(2.69)	(4.24)	(3.43)	(0.93)
$\ln(A)^2$			−0.065**			−0.064*			−0.096**
			(−2.47)			(−1.93)			(−2.06)
$\ln(L)^2$			−0.038*			−0.031			−0.065
			(−1.85)			(−1.19)			(−0.72)
$\ln(L) \times \ln(A)$			0.065***			0.069			0.013
			(1.71)			(1.45)			(0.13)
L/A		−0.0001***			-7×10^{-5}**			-3×10^{-5}	
		(−2.82)			(−2.39)			(−0.82)	
M	3×10^{-5}*	3×10^{-5}*	3×10^{-5}**	8×10^{-5}**	8×10^{-5}**	8×10^{-5}**	2×10^{-5}	2×10^{-5}***	2×10^{-5}***
	(1.84)	(1.86)	(1.99)	(2.00)	(2.08)	(2.06)	(2.77)	(2.82)	(3.18)
G	1×10^{-5}	1×10^{-5}	2×10^{-5}	-1×10^{-5}	-1×10^{-5}	-1×10^{-5}	2×10^{-6}*	2×10^{-5}*	3×10^{-5}**
	(0.47)	(0.48)	(0.75)	(−0.16)	(−0.2)	(−0.17)	(1.66)	(1.74)	(2.26)
C	2×10^{-5}	2×10^{-5}	3×10^{-5}	2×10^{-5}	1×10^{-5}	2×10^{-5}	3×10^{-5}**	3×10^{-5}***	3×10^{-5}***
	(1.15)	(1.14)	(1.38)	(0.40)	(0.3)	(0.42)	(2.58)	(2.6)	(3.00)
farm size	0.008*	0.008*	0.01*	0.002	0.002	0.003	0.014**	0.014**	0.016***
	(1.68)	(1.70)	(1.95)	(0.36)	(0.36)	(0.55)	(2.21)	(2.22)	(2.61)
Ext	0.026	0.024	0.025	−0.009	−0.01	−0.01	0.134***	0.135***	0.134***
	(1.2)	(1.14)	(1.18)	(−0.35)	(−0.36)	(−0.38)	(4.83)	(4.82)	(4.68)
plotage	0.017***	0.017***	0.018***	0.028***	0.027***	0.028***	0.0153***	0.0148***	0.016***
	(3.2)	(3.21)	(3.38)	(3.00)	(2.82)	(2.87)	(3.68)	(3.55)	(3.79)
plotage^2	−0.0001	-9×10^{-5}	-9×10^{-5}*	−0.0003**	−0.0002*	−0.002*	-7×10^{-5}**	-7×10^{-5}*	−0.0001*
	(−1.56)	(1.54)	(−1.66)	(−2.03)	(−1.71)	(1.69)	(−1.77)	(−1.72)	(−1.86)
Sigatoka	−0.21	−0.204	−0.214	−0.183	−0.167	−0.17	−0.737	−0.742	−0.744
	(−1.59)	(−1.56)	(−1.64)	(−1.18)	(−1.08)	(−1.09)	(−1.59)	(1.6)	(−1.61)
weevils	−0.035	−0.063	−0.088	0.049	0.012	−0.0009	−0.161	−0.153	−0.172*
	(−0.34)	(−0.60)	(−0.83)	(0.36)	(0.09)	(−0.01)	(−1.53)	(1.46)	(−1.68)
High elevation	0.54***	0.52***	0.523***						
	(5.12)	(4.98)	(5.23)						
Log likelihood	−624.9	−621.2	−620.6	−500.1	−497.4	−497.7	−51.4	−51.1	−48.2
TE	0.475	0.478	0.474	0.437	0.442	0.44	0.705	0.703	0.706

Notes: ***, **, * indicate statistical significance at the 1, 5, and 10 percent levels, respectively. Eql is the Cobb-Douglas technology, Eq2 is the transcendental production function, and Eq3 the transcendental logarithmic function (translog). See Table 8.1 for definitions of the variables.

extension agents could enable farmers to adopt new farming techniques and to raise their production frontier. However, it is possible that the extension agents visit the most productive farmers, without necessarily increasing farmers' use of new technologies and practices.

The effect of age of a banana plot is significant (positive for young plots and negative for old ones) for all the cases. Age of the banana plot was included in the estimation to account for the low yields observed in young plantations and old ones. Results for black Sigatoka show that it has a negative effect on banana production, but the effect is not significant. The effect of weevils is also not significant. The insignificant results obtained for black Sigatoka and weevils for the overall sample might be due to correlation between the disease/pest and the location dummy (elevation). Excluding the location dummy variable from the estimation makes the coefficients of the black Sigatoka score and weevil score significant at 5 percent and 10 percent, respectively (Table 8.3).

The regional dummy variable (location in the high-elevation region) has a positive and significant effect (1 percent) on output. This result shows that elevation-related effects are important in determining the productivity of bananas. The dummy was included in the equations for the overall sample to capture regional differences and to capture the differences in biotic and abiotic factors characterizing the two different regions.

The TE scores reveal the presence of inefficiency especially for low-elevation region. The TE score obtained for the high-elevation was higher than that for the low-elevation region, implying that inefficiency contributes to the low output realized in the region. The TE scores obtained by using different function forms were very close, implying that model specifications for the frontier function have no impact on the predicted technical efficiencies for the sample farmers. This finding is consistent with what is reported in the literature (Kopp and Smith 1980).

Table 8.4 shows elasticities of production, with respect to labor and land, and the returns to scale for cooking bananas production in the two regions. The elasticity of labor is higher for the low-elevation region, while the opposite is true for crop area. The sum of the elasticities of labor and land are all below 1 in all the cases, which indicates decreasing returns to scale. The implication of this result is that farmers would loose efficiency if they increase scale of production. The decrease in efficiency as a result of the increase in scale of production is most likely due to differences in soil quality between small and large plots. This result is consistent with that obtained for farm size. Given farm size, increasing plot size leads to a decrease in banana productivity.

The three functional forms (Cobb-Douglas, transcendental, and translog) yield different results in terms of elasticities. The elasticities of labor obtained from the transcendental function are much higher than those obtained from the Cobb-Douglas and translog forms. However, the returns to scale obtained from all functions are quite close. The Cobb-Douglas seems to be a consistent and appropriate function for assessing production technology in the two different regions.

The data support rejection of the null hypothesis that farmers growing cooking bananas are technically efficient in all cases (Table 8.5). The results for factors influencing TE are shown in Table 8.6.

The effect of age on TE is not significant, perhaps as a consequence of multicollinearity. However, the relationship between age and TE is not as expected. TE first decreases as age increases in the early years but later starts to increase, as shown by the negative effect of the quadratic term. The reason for this behavior could be associated with the reproduction process of the household. Young families may allocate more of their time to raising children, so that more time is available for farm production as the

Table 8.3 Cobb-Douglas production estimates for the overall sample (location dummies excluded)

Variable	Coefficient	*t*-value
Production function estimates		
Constant	5.773***	18.55
ln(*A*)	0.34***	8.17
ln(*L*)	0.422***	9.66
M	0.00003*	1.65
G	0.00002	0.88
C	0.00002	1.28
farm size	0.006	1.32
Ext	0.017	0.74
plotage	0.024***	4.62
$plotage^2$	–0.0001**	–2.10
Sigatoka	–0.304**	–2.36
weevils	–0.167*	–1.65
Log likelihood	–639.1	
Wald X^2	726.7	
TE	0.449	
Technical inefficiency estimates		
Constant	0.511	0.71
Age	0.024	0.82
Age^2	–0.0002	–0.76
Hplot	–0.533***	–3.13
Edhh	0.018	1.02
Hhsz	–0.053*	–1.83
depr	0.368	1.26
Kk	–0.001	–1.00
D	–0.005	–0.84
σ_V (standard error)	0.38	0.048

Notes: ***, **, and * indicate statistical significance at the 1, 5, and 10 percent levels, respectively. See Table 8.1 for definitions of the variables.

children grow up. The children also contribute labor, enabling the farmers to implement timely management decisions.

The education variable gives mixed results, as expected. In the low-elevation region, the impact of education on TE is negative. This finding is consistent with our hypothesis that educated households are less efficient if education increases the opportunity cost of labor, so that farmers reallocate resources from farm to nonfarm activities. The impact of education on TE in the high-elevation region is positive. In this region, education appears to increase farmers' management capabilities and their ability to utilize technologies.

The results for the relationship between the distance variable and TE show a negative relationship but is only significant for high elevations. Distance to paved roads is

Table 8.4 Production elasticities

Region	Elasticities of production: Labor	Land	Returns to scale
Overall sample			
Cobb-Douglas	0.384	0.332	0.716
Transcendental	0.489	0.207	0.696
Translog	0.343	0.314	0.658
Low elevation			
Cobb-Douglas	0.395	0.332	0.727
Transcendental	0.5	0.206	0.706
Translog	0.377	0.316	0.693
High elevation			
Cobb-Douglas	0.282	0.264	0.546
Transcendental	0.343	0.199	0.542
Translog	0.296	0.261	0.557

Table 8.5 Test for the null hypothesis that $\sigma_u = 0$

Region	χ^2	P	Outcome
Overall sample	45.08	0.000	Reject null
Low elevation	29.52	0.000	Reject null
High elevation	8.9	0.004	Reject null

positively correlated with TE. The distant farms are more technically efficient, most likely because of access to cheap labor, which enables them to implement timely management decisions.

The family size variable is positively related to TE but is only significant, at 10 percent, for the whole sample and the low-elevation region. Households with big families are more technically efficient, most likely because they strive to achieve higher output to meet their subsistence requirements. Moreover, large families have more labor endowment (including children) needed to implement management decisions. The impact of the dependence ratio on TE is negative but only significant for the high-elevation region.

The gender of the plot manager (husband = 1 and wife = 0) has a positive and significant impact on TE for the whole sample (1 percent) and for the low-elevation region (5 percent). In the high-elevation region, the relationship is negative but not significant. Higher efficiency in plots managed by husbands can be explained by differential access to production resources and hence timing of input application and management practices.

Results suggest that access to credit improves efficiency in production of cooking bananas in all cases, but the effect is not significant. This result implies that liquidity constraints affect farmers' ability to apply inputs and implement farm management decisions on time.

Table 8.6 Factors influencing technical efficiency

	Overall sample			Low elevation			High elevation		
Variable	Eql	Eq2	Eq3	Eql	Eq2	Eq3	Eql	Eq2	Eq3
Constant	0.415	0.28	0.263	0.438	0.263	0.267	−2.375*	−2.312	−2.358
	(0.54)	(0.36)	(0.35)	(0.50)	(0.29)	(0.30)	(1.3)	(−1.28)	(−1.33)
Age	0.02	0.027	0.025	0.033	0.04	0.039	0.039	0.04	0.042
	(0.67)	(0.86)	(0.83)	(0.95)	(1.13)	(1.09)	(0.51)	(0.53)	(0.59)
Age^2	−0.002	−0.0003	−0.0003	−0.0004	−0.004	−0.0004	−0.0003	−0.0003	−0.0004
	(−0.70)	(−0.89)	(−0.83)	(−1.05)	(−1.22)	(−1.17)	(0.43)	(−0.46)	(−0.5)
edhh	0.016	0.015	0.019	0.013	0.013	0.015	−0.037	−0.043	−0.044
	(0.9)	(0.85)	(1.03)	(0.62)	(0.62)	(0.70)	(−0.86)	(−0.99)	(−1.00)
D	−0.001	−0.002	0.0008	−0.006	−0.006	−0.005	−0.064***	−0.065***	−0.061***
	(−0.22)	(−0.28)	(0.13)	(−0.95)	(−0.97)	(−0.70)	(−2.88)	(−2.98)	(−2.79)
hhsz	−0.054*	−0.056*	−0.055*	−0.059	−0.062*	−0.061*	−0.049	−0.047	−0.032
	(−1.78)	(−1.86)	(−1.86)	(−1.65)	(−1.73)	(−1.71)	(−0.57)	(0.55)	(−0.39)
depr	0.443	0.436	0.436	0.506	0.506	0.504	1.505*	1.513*	1.378*
	(1.42)	(1.40)	(1.42)	(1.39)	(1.38)	(1.39)	(1.73)	(1.78)	(1.66)
hplot	−0.567***	−0.579***	−0.577***	−0.522**	−0.535**	−0.53**	0.293	0.288	0.184
	(−3.21)	(−3.27)	(−3.29)	(−2.54)	(−2.59)	(−2.58)	(0.63)	(0.63)	(0.41)
kk	−0.001	−0.001	−0.001	−0.001	−0.001	−0.001	−0.015	−0.018	−0.022
	(−1.04)	(−1.04)	(−1.03)	(−1.28)	(−1.28)	(−1.27)	(−1.04)	(1.06)	(−0.74)
sk	0.0002	0.0002	0.0002	0.0004	0.0004	0.0004			
	(1.15)	(1.12)	(1.15)	(1.40)	(1.36)	(1.4)			
σ_V (standard error)	0.429	0.43	0.414	0.483	0.491	0.483	0.205	0.201	0.192
	(0.049)	(0.05)	(0.05)	(0.069)	(0.071)	(0.074)	(0.032)	(0.032)	(0.035)

Notes: ***, **, * indicate statistical significance at the 1, 5, and 10 percent levels, respectively. Eql is the Cobb-Douglas technology, Eq2 is the transcendental production function, and Eq3 is the transcendental logarithmic production function (translog). See Table 8.1 for definitions of variables. A positive sign on a coefficient implies a negative effect on efficiency and vice versa.

The results on the interaction between SOM and K, and physical (sand) and chemical (pH) characteristics and the effect on productivity are presented in Table 8A.3. The estimates from a three-stage least squares method show that the proportion of sand in the soil negatively affects SOM content. The results also show that the SOM content is higher in Masaka, implying that differences in regional characteristics affect SOM accumulation and decomposition. It should be noted that SOM is highly correlated with N content in the soil. Availability of K is positively influenced by the SOM content in the soil, pH, and additions of crop residues. In turn, K availability positively affects the yield of cooking bananas, as expected, but the effect is not significant. However, the effect of SOM on cooking-banana yield is negative, but only significant at 10 percent. This result can be explained by the conditions that favor accumulation of SOM, but are not favorable for the production of cooking bananas. SOM tends to accumulate faster in clay soils, which are not good for production of cooking bananas because of physical impediment of banana root growth. Another reason could be related to the C:N ratio of materials used in the formation of the SOM. SOM with high C:N ratios can affect availability of nutrients through immobilization of the nutrients during the SOM decomposition. Animal manure has a positive and significant (10 percent) effect on yields of cooking bananas. The effect of plot age is significant at 1 percent (positive for young plots and negative for older plots). The effect of black Sigatoka is negative and significant at 5 percent.

Finally, we estimate the reduced form using the frontier function approach (Table 8.7). The elasticities of labor and crop area are positive, as expected. The sum of the elasticities, from the Cobb-Douglas function, indicates constant returns to scale. This result contrasts with the result obtained for the main sample, which displays decreasing returns to scale. Most likely the case study sites are not representative of the main sample; hence the difference in the results obtained for returns to scale. However, the case study results sheds some light on the contribution of biophysical characteristics to the shift of banana production from the low- to the high-elevation region.

Animal manure has a positive effect on productivity, with a statistical significance of 1 percent. The effect of sand on the productivity of cooking bananas is positive and significant at 1 percent. The effect of pH is positive and significant at 5 percent for all the model specifications. The average TE obtained (44.9 percent to 45.6 percent, depending on function form) from the case study is close to those obtained for the main sample.

Conclusions

One of the objectives of this research report is to assess the impact of improved banana technology on smallholder farmers in the lake region of Uganda and Tanzania. In this chapter, we use the stochastic production functions to analyze productivity and efficiency of smallholder farmers in Uganda. In the chapter, we assess the impact of plot, farm, and regional characteristics on the productivity and efficiency of banana farmers. We also analyze the impact of soil organic amendments on banana productivity. Findings show that to improve the productivity of small farmers, much more will need to be done in terms of access to basic inputs, farm credit, information, and education.

The productivity of cooking bananas depends on the climate and soil characteristics of regions, and, as hypothesized, it is higher in the high-elevation region. Labor and crop area respond positively to output, but the scale elasticity is below 1 in all cases, implying that farmers cannot increase scale of production without losing efficiency. Labor productivity is higher in the low-elevation region, where most agronomic practices (such as crop sanitation) are carried out minimally. Animal manure, grass mulch, and crop residues have a positive and significant effect on productivity, especially in the high-elevation region. Soil pH has a positive and significant effect on productivity, whereas the effect of soil organic matter and soil texture (sand) is not significant. For a given banana plot size, banana yields increase with farm size. Alternatively, keeping farm size constant, an increase in plot size decreases the productivity of bananas. Extension visits enhance productivity in the high- but not in the low-elevation region.

Findings illustrate substantial inefficiencies in the production of cooking bananas, especially in the low-elevation region. Education improves technical efficiency in the highlands but not in the low-elevation region. Market access (using distance to paved road as a proxy) reduces efficiency, especially for farmers in the high-elevation region. Household size is positively related to efficiency. Banana production appears to be more efficient when managed by men than by women, probably because of underlying differentials in access to resources. Access to credit increases efficiency, whereas rent and remittances reduce efficiency.

Policies to improve production efficiency include investments in education and extension services, and improving access to production credit. Clearly, policies will need to be tailored to the local conditions, as demonstrated by the different production and efficiency profiles of the two regions analyzed here.

Table 8.7 Incorporating soil factors in the production frontier function (n = 154)

	Cobb-Douglas		Transcendental		Translog	
Variable	Coefficient	*z*-value	Coefficient	*z*-value	Coefficient	*z*-value
Stochastic frontier function						
Constant	2.612***	2.75	2.533**	2.35	4.509	1.56
$\ln(A)$	0.446***	4.52	0.426**	2.56	0.624	0.84
$\ln(L)$	0.549***	6.79	0.568***	3.88	–0.18	–0.17
$\ln(A)^2$					–0.048	–0.59
$\ln(L)^2$					0.062	0.71
$\ln(L) \times \ln(A)$					–0.049	–0.4
L/A			–0.00002	–0.15		
M	0.0001***	3.8	0.0001***	3.69	0.0001***	3.27
C	0.00004	0.44	0.00004	0.45	0.00001	0.13
sand	0.02***	3.42	0.245***	3.39	0.264***	3.38
pH	0.245**	2.43	0.02**	2.44	0.021**	2.46
plotage	0.033***	3.89	0.033***	3.78	0.034***	3.86
$plotage^2$	–0.0003***	-3.27	–0.0003***	-3.21	–0.0003***	-3.28
Sigatoka	–1.5***	-3.83	–1.511***	-3.8	–1.517***	-3.79
Log likelihood	-193.8		-193.7		-193.2	
TE (standard deviation)	0.451		0.449		0.456	
Factors influencing technical inefficiency						
Constant	0.453	0.34	0.456	0.34	0.304	0.22
Age	–0.001	–0.02	–0.001	–0.02	0.007	0.12
Age^2	0.0001	0.23	0.0001	0.22	0.00005	0.09
hplot	–0.355	–0.94	–0.35	–0.92	–0.399	-1.03
edhh	–0.017	–0.45	–0.017	–0.45	–0.018	–0.48
hhsz	0.044	0.89	0.045	0.9	0.035	0.68
depr	–0.096	–0.17	–0.094	–0.17	–0.051	–0.09
σ_V (standard error)	0.383	0.1	0.378	0.105	0.395	0.22

Notes: ***, **, and * indicate statistical significance at the 1, 5, and 10 percent levels, respectively. TE is technical efficiency. A blank entry indicates that the variable was not included in the regression. See Table 8.1 for definitions of the variables.

Supplementary Tables

Table 8A.1 Labor demand estimates (first stage of the production function estimation)

	Overall sample		Low elevation		High elevation	
Variable	Coefficient	*t*-value	Coefficient	*t*-value	Coefficient	*t*-value
Constant	6.425***	16.19	6.247***	12.59	6.808***	15.62
ln(*A*)	0.513***	13.62	0.498***	10.73	0.36***	6.48
ln(*w/p*)	−0.397***	-3.93	−0.46***	-3.82	−0.296*	-1.7
D	−0.009***	-3.94	−0.01***	-3.90	−0.06	-1.24
hhsz	0.027	1.57	0.03	1.37	0.052**	2.52
depr	−0.453***	-2.61	−0.58***	-2.64	0.018	0.09
Age	−0.005	−0.34	−0.008	−0.40	−0.005	−0.27
Age^2	0.0001	0.72	0.0002	0.92	0.0001	0.43
hplot	0.264***	2.8	0.314***	2.68	0.164	1.32
edhh	0.025**	2.51	0.034***	2.79	0.009	0.83
plotage	0.025***	4.52	0.018*	1.78	0.009	1.46
$plotage^2$	−0.0002***	-3.2	−0.0002	−1.38	−0.0001	−1.53
Sigatoka	−4.22***	-2.68	−0.331*	−1.88	−0.578	−0.96
weevils	−0.005	-0.04	0.266*	1.74	−0.261*	−1.85
Adjusted R^2	0.482		0.404		0.408	

Notes: ***, **, and * indicate statistical significance at the 1, 5, and 10 percent levels. See Table 8.1 for definitions of the variables.

Table 8A.2 Two-stage least squares estimates of the production function for cooking bananas (endogeneity test)

	Overall		Low elevation		High elevation	
Variable	Coefficient	*t*-value	Coefficient	*t*-value	Coefficient	*t*-value
Constant	6.44***	7.48	6.229***	7.39	5.426***	3.37
ln(*A*)	0.532***	6.53	0.516***	6.10	0.256**	2.48
ln(*L*)	0.104	0.76	0.112	0.81	0.427*	1.8
M	0.00003*	1.77	0.0001**	2.28	0.00002**	2.32
G	0.00001	0.41	0.00006	0.76	0.000001	0.48
C	0.00003	1.17	0.00003	0.59	0.00003*	1.83
farm size	0.01**	2.03	0.003	0.52	0.018***	2.6
Ext	0.016	0.76	−0.0001	−0.01	0.124***	4.78
plotage	0.032***	4.36	0.039***	3.59	0.005	0.92
plotage^2	−0.0002***	−2.78	−0.0003**	−2.00	0.00002	0.31
Sigatoka	−0.096	−0.58	0.015	0.08	−0.542	−0.98
weevils	−0.108	−0.85	0.01	0.06	−0.114	−0.86
High elevation	0.715***	5.99				
Residual[a]	0.384**	2.60	0.383**	2.57	−0.145	−0.58
Adjusted R^2	0.662		0.491		0.687	

Notes: ***, **, and * indicate statistical significance at the 1, 5, and 10 percent levels, respectively. See Table 8.1 for definitions of the variables. A blank entry indicates that the variable was not included in the regression. See Table 8.1 for definitions of the variables.

[a]Residual is from the first-stage estimation (the labor equation); variables included in the labor equation are area under cooking bananas, wage/price ratio, distance to paved roads, household characteristics (size, composition, age education, and gender), and village characteristics representing opportunities (farm wage income, nonfarm wage income, salary income, and self-employment earnings).

Table 8A.3 Production function estimates, three-stage least squares estimate

	ln(*SOM*)		ln(*K*)		ln(*Y*)	
Variable	**Coefficient**	***t*-value**	**Coefficient**	***t*-value**	**Coefficient**	***t*-value**
Constant	2.37***	18.09	–4.498***	–13.73	5.601***	3.58
ln(*A*)					0.32***	2.77
ln(*L*)					0.679***	6.83
M					9.0×10^{-5}***	2.76
C			7.0×10^{-5}**	2.29		
ln(*SOM*)			1.135***	4.06	–1.479	–1.54
ln(*K*)					0.415	1.05
sand	–0.014***	–6.23				
pH			0.442***	4.99		
plotage					0.035***	3.48
plotage^2					–0.0003***	–2.82
Sigatoka					–0.971**	–2.18
High elevation	–0.058	–1.23				
Adjusted R^2	0.23		0.59		0.59	

Notes: *** and ** indicate statistical significance at the 1, 5, and 10 percent levels, respectively. The *sand* variable instruments *SOM*, and *pH* instruments *K*. See Table 8.1 for definitions of the variables.

References

Aigner, D. J., C. A. K. Lovell, and P. Schmidt. 1977. Formulation and estimation of stochastic frontier production function models. *Journal of Econometrics* 6 (1): 21–37.

Ali, M., and D. Byerlee. 1991. Economic efficiency of small farmers in a changing world: A survey of recent evidence. *Journal of International Development* 3 (1): 1–27.

Battese, G. E., and S. S. Broca. 1997. Functional forms of stochastic frontier production functions and models for technical inefficiency effects: A comparative study for wheat farmers in Pakistan. *Journal of Productivity Analysis* 8 (4): 395–414.

Battese, G. E., and T. J. Coelli. 1995. A model for technical inefficiency effects in a stochastic frontier production function for panel data. *Empirical Economics* 20 (2): 325–332.

Bauer, P. W. 1990. Recent development in econometric estimation of frontiers. *Journal of Econometrics* 46 (1–2): 39–56.

Belbase, K., and R. Grabowski. Technical efficiency in Nepalese agriculture. *Journal of Development Areas* 19 (4): 515–525.

Binam, J. N., K. Sylla, I. Diarra, and G. Nyambi. 2003. Factors affecting technical efficiency among coffee farmers in Cote d'Ivoire: Evidence from the centre west region. *R&D Management* 15 (1): 66–76.

Bravo-Ureta, B. E., and R. E. Evenson. 1994. Efficiency in agricultural production: The case of peasant farmers in Eastern Paraguay. *Agricultural Economics* 10 (1): 27–37.

Bravo-Ureta, B. E., and E. Pinheiro. 1997. Technical, economic and allocative efficiency in peasant farming: Evidence from the Dominican Republic. *Developing Economies* 35 (1): 48–67.

Coelli, T. J. 1994. A guide to Frontier 4.1: A computer program for stochastic frontier production and cost function estimation. Department of Econometrics, University of England, Armidale, Australia. Mimeo.

———. 1995. Recent developments in the frontier modeling and efficiency measurement. *Australian Journal of Agricultural Economics* 39 (3): 219–245.

Coelli, T. J., D. S. P. Rao, and G. E. Battese. 1998. *An introduction to efficiency and productivity analysis.* Amsterdam: Kluwer Academic Publishers.

Farrell, M. J. 1957. The measurement of productive efficiency. *Journal of the Royal Statistical Society,* Series A 120 (3): 253–290.

Forsund, F. R., C. A. K. Lovell, and P. Schmidt. 1980. A survey of frontier production functions and their relationships to efficiency measurement. *Journal of Econometrics* 13 (1): 5–25.

Fried, H. O., C. A. K. Lovell, and S. S. Schmidt. 1993. *The measurement of productive efficiency: Techniques and applications.* Oxford: Oxford University Press.

Hussain, S. S. 1989. Analysis of economic efficiency in Northern Pakistan: Estimation, causes and policy implications. Ph.D. dissertation, University of Illinois, Urbana–Champaign.

Kalirajan, K. P. 1981. An econometric analysis of yield variability in paddy production. *Canadian Journal of Agricultural Economics* 29 (2): 283–294.

———. 1984. Farm-specific technical efficiencies and development policies. *Journal of Economic Studies* 11 (3): 3–13.

———. 1991. The importance of efficiency use in adoption of technology: A micro panel data analysis. *Journal of Productivity Analysis* 2 (2): 113–126.

Kalirajan, K. P., and R. T. Shand. 1986. Estimating location specific and farm specific technical efficiency: An analysis of Malaysian agriculture. *Journal of Economic Development* 11 (2): 147–160.

———. 1997. Modeling and measuring technical efficiency: An alternative approach. In *Issues in agricultural competitiveness: Markets and policies,* ed. R. Rose, C. Tanner, and M. A. Bellamy. Proceedings of the 22nd International Conference of Agricultural Economists, August 22–29, 1994, Harare, Zimbabwe. Hants, U.K.: Dartmouth Publishing.

Kopp, R. J., and V. K. Smith. 1980. Frontier production function estimates for steam electric generation: A comparative analysis. *Southern Economics Journal* 47 (1): 1049–1059.

Kumbhakar, S. C., S. Gosh, and J. T. McGuckin. 1991. A generalized production frontier approach for estimating determinants of inefficiency in U.S. dairy farms. *Journal of Business and Economic Statistics* 9: 279–286.

Maddala, G. S. 1971. The use of variance models in pooling cross section and time series data. *Econometrica* 39 (2): 341–358.

Meeusen, W., and van den Broeck, J. 1977. Efficiency estimation from Cobb-Douglas production functions with composed error. *International Economic Review* 18 (4): 435–444.

Olsen, S. R., C. V. Cole, F. S. Watanabe, and L. A. Dean. 1954. Estimation of available phosphorus in soils by extraction with sodium bicarbonate. U.S. Department of Agriculture Circular 939. Washington, D.C.: U.S. Government Printing Office.

Phillips, M. J., and R. P. Marble. 1986. Farmer education and efficiency: A frontier production function approach. *Economy of Education Review* 5: 257–264.

Pitt, M. M., and L. F. Lee. 1981. The measurement and sources of technical inefficiency in the Indonesian weaving industry. *Journal of Development Economics* 9 (1): 43–64.

Rufino, M. 2003. On-farm analysis of nematode infestation and soil fertility as constraints to the productivity of banana-based production systems in Uganda. M.S. thesis, Plant Sciences Department, Production Systems Group, Wageningen University, The Netherlands.

Schmidt, P. 1986. Frontier production functions. *Econometric Review* 4 (2): 329–334.

Thomas, G. W. 1982. Exchangeable cations. In *Methods of soil analysis*. Part 2: *Chemical and biological properties*, ed. A. L. Page, R. H. Miller, and D. R. Keeney. Madison, Wis., U.S.A.: American Society of Agronomy and Soil Science Society of America.

Tian, W., and G. H. Wan. 2000. Technical efficiency and its determinants in China's grain production. *Journal of Productivity Analysis* 13 (2): 159–174.

Tisdale, S. L., W. L. Nelson, J. D. Beaton, and J. L. Havlin. 1993. *Soil fertility and fertilizers*. Fifth edition: Upper Saddle River, N.J.: Prentice-Hall.

Weir, S. 1999. *The effects of education and farmer productivity in rural Ethiopia*. Working Paper CSAE WPS99-7. Oxford: Centre for the Study of African Economics, University of Oxford.

Weir, S., and J. Knight. 2000. *Education externalities in rural Ethiopia: Evidence from average and stochastic production functions*. Working Paper CSAE WWPS2000-4. Oxford: Centre for the Study of African Economics, University of Oxford.

CHAPTER 9

Use of Hybrid Cultivars in Kagera Region, Tanzania, and Their Impact

Svetlana Edmeades, Jackson M. Nkuba, and Melinda Smale

As described in Chapter 5, the formal introduction of new banana cultivars to Kagera Region began in 1997 with KCDP. Preliminary studies by the project indicate rapid and extensive uptake of these materials. This chapter re-examines the use of introduced banana hybrids and one aspect of their social and economic impact: reduced vulnerability to production losses from biotic pressures. A treatment model is used to identify the determinants of adoption and the effects of adoption on expected yield losses from pests and diseases, while controlling for other factors. Findings demonstrate that the intended impact of growing hybrids on production vulnerability was achieved, which lends justification to research efforts aimed at developing resistance materials. Like other analyses in this report, this study shows the preeminence of farmer-based systems for diffusing planting material, as compared to formal systems.

The agriculture of Kagera Region, located in the northwestern corner of Tanzania on the western shores of Lake Victoria, is based largely on banana/coffee farming systems.[1] About 43.8 percent of the total acreage under banana production in Tanzania is found in this region. Over the past decades, Kagera Region has experienced serious decline in the production of bananas and other traditional crops, including coffee, because of declining soil fertility and increasing infestations of pests and diseases (Bosch et al. 1996; Baijukya and Folmer 1999). The situation has also been made worse by the persistence of outdated methods of crop husbandry practiced by banana growers and the lack of reliable local and external markets for bananas, leading to frequent periods of food insecurity.

As detailed in Chapter 4, banana growers, national and international research institutions, and NGOs have sought to formulate strategies to mitigate production decreases. One of these strategies was the introduction of new banana cultivars into the region. Since the 1960s, several exotic cultivars have been introduced by farmers into the region from other banana growing areas within and outside the country. The formal introduction of new banana cultivars into the region began in 1997, when KCDP was established. One of its objectives was to acquire, multiply, and disseminate new banana cultivars to farmers.

On-farm testing of the new cultivars commenced in 1997 and was conducted concurrently with their multiplication and dissemination of planting materials to farmers in the region

[1] This chapter is drawn from a selected paper presented at the international conference Economics of Poverty, Environment, and Natural Resource Use, May 2006, Wageningen University, The Netherlands.

(Nkuba et al. 1999). The results of on-farm testing showed that on average the new banana cultivars yielded a bunch weight of 18.9 kg, compared with 9.7 kg for local cultivars (Byabachwezi, Steenhuijesen, and Rwezaula 1997; Nkuba, Ndege, and Mkulila 2002). Initial assessment revealed that these cultivars are acceptable to farmers for their multiple uses (cooking, dessert, roasting, and brewing) and good marketability. From 1997 to 2002 about 2.5 million suckers (or planting material) from the new banana cultivars had already been distributed to farmers. Knowledge gaps remain, however. For example, although KCDP analyzes initial adoption rates and acceptability in project reports, more systematic information about the extent and determinants of adoption, as well as the economic impacts of adoption, is needed.

In Tanzania, there is increasing need for investment in the improvement of crops and livestock, which is one of the government strategies on reduction of rural poverty (MAFS 2001). Similar to other developing economies, despite the large financial resources spent on development programs each year, there is limited awareness of the actual impact of programs on the poor (Baker 2000). This situation poses basic questions about whether targeted interventions have achieved the intended benefits and their effectiveness and efficiency. Although banana growers in Tanzania have participated in a number of extension and research programs, there is little documentation of program impact.

This chapter focuses on identifying the determinants of use of the new banana cultivars and assessing the impacts of use on production vulnerability. The new banana cultivars are the hybrids (from FHIA) formally introduced by KCDP in the region. The analysis is used to better understand the characteristics and production potential of adopters compared to nonadopters in areas exposed to the program interventions. Other aspects of adoption and impact are analyzed in Nkuba (2007).

The next section describes household income-earning strategies and the resource base of household farms in the banana-based production system of Kagera Region. The conceptual approach is the theoretical framework of the agricultural household, presented in Chapter 2. The econometric approach is a treatment effects model, described below. The model is applied to data collected through a stratified random sample of households (Chapter 2 and Appendix D). Stratifying variables include exposure to new cultivars through programs, to control for treatment effects, and elevation. Elevation is related to pest and disease pressures. We contend that pest and disease pressures have driven farmers to seek new, disease-free planting material.

Production Systems and Income Sources

The banana/coffee-based farming system is dominated by smallholder farmers, with an average farm size ranging from 0.5 to 1.5 ha. There are three types of land use, locally known as *Kibanja, Kikamba,* and *Rweya. Kibanja* is a permanent field for cultivation of bananas. Often, bananas are intercropped with coffee and, during the rainy season, with beans. Although households sell surplus banana production for cash generation, coffee is the main cash crop, with the region leading in coffee production in the country. Tea is also grown along the lake shores of the region. Other crops intercropped in *Kibanja* are maize, and root and tuber crops, fruits, and vegetables. *Kibanja* provides food and income for the household. *Kikamba* is an area found adjacent to *Kibanja,* and it is used for cultivating annual crops (beans, maize, and groundnuts). *Rweya* is grassland used for grazing, cutting mulch grass, and shifting cultivation of annual crops, such as bambara nuts (Lorkeers 1995).

Among the banana cultivars grown in Kagera Region are the EAHBs, which have been endemic (or constantly present) in the

Table 9.1 Five-year average areas under crop production by district in Kagera Region, 1996/97–2000/01 (ha)

Crop type	Bukoba	Muleba	Biharamulo	Ngara	Karagwe	Total
Foodcrops						
Bananas	47,830	36,807	1,870	25,549	41,289	153,345
Maize	11,490	4,440	25,445	4,347	25,523	71,245
Sorghum	—	—	8,675	3,056	2,500	14,231
Sweet potatoes	15,970	6,060	13,647	151	6,961	42,669
Cassava	15,551	15,401	33,570	9,524	34,763	108,809
Paddy	—	—	1,826	—	—	1,826
Beans	21,560	8,631	16,795	5,786	32,856	85,628
Potatoes	—	—	—	—	361	361
Subtotal	112,401	71,339	101,829	48,292	144,254	478,115
Cash crops						
Coffee	31,434	19,374	617	762	15,895	68,082
Tea	1,129	133	—	—	—	1,262
Cotton	—	—	7,191	—	—	7,191
Subtotal	32,563	19,507	7,808	762	15,895	76,535
Total	144,964	90,846	109,637	49,054	160,149	554,650

Source: Regional Commissioner's Office, Bukoba, Kagera Region, 2003.
Note: — indicates none reported.

region since 500–1000 A.D. (Byabachwezi and Mbwana 1999), and other exotic and hybrid bananas. Based on their uses, bananas produced in the region can be classified into four use types: cooking, beer, roasting, and sweet (dessert) (see Chapter 3).

Bananas cover about 33 percent and 26 percent of the total land under crops in Bukoba and Karagwe districts, respectively, of Kagera Region, where this study was implemented (Table 9.1). During 1996–2001, annual average production of bananas was 919,995 tons (RCO 2003). Karagwe District contributed about 45 percent of the total production in the region, with the Bukoba contributing 23 percent, Muleba 20 percent, Ngara 11 percent, and Biharamulo 1 percent. Banana production is seasonal, with the peak season spanning between the months of June and August, and the lowest production levels attained during November.

Measuring Impacts

The literature about economic impacts of program intervention emphasizes the importance of establishing the appropriate counterfactual. As stated by Ravallion (1994, 4), "the essential problem of impact evaluation is that we do not observe the outcomes for participants if they had not participated." The appropriate counterfactual facilitates measurement of the correct causal relationship between the technology and the outcomes being measured, because other confounding factors may also have influenced the outcome (Ravallion 1994; Baker 2000; Doss 2003; Chapter 2).

Ravallion (1994) summarizes and compares the five main methods available for evaluating program impact. Using randomization, individuals are selected into treatment and comparison groups at random, so that the only measurement errors are associated with sampling. Sampling errors can be reduced through larger samples. In the matching approach, the comparison group is matched to the treatment group based on characteristics measured in data from a larger, representative sample survey. Propensity scores are tabulated to support the selection of individuals. Reflexive comparisons enable the "before and after" to be estimated for key parameters using a baseline, which serves as the comparison group. With the double difference method, the treatment and comparison group are compared both before and after the treatment. The fifth approach is the instrumental variables approach used in this chapter. Instrumental variables are those that matter to participation, but not to the outcome. They enable the identification of exogenous variation in outcomes that can be attributed to the program.

In our case, the participation variable is the decision to use new banana hybrids. The econometric approach for measuring the impact of using these new banana hybrids is described next.

Econometric Approach

Use of banana hybrids and their impact on smallholder farmers are tested with a treatment effects model (defined in Maddala 1983). The model is designed to capture the effect of an endogenously chosen binary treatment (use of banana hybrids) on another continuous observed variable that expresses a social or economic impact on households. The approach controls for the potential endogeneity of hybrid use decisions and outcomes through the use of two stratification variables: elevation (correlated with biotic pressures) and exposure of the surrounding area to program interventions. Stratification variables can be used as instruments for the identification of one of the behavioral outcomes—in this case, hybrid use. In one equation of the simultaneous system, the determinants of hybrid use are identified. The factors influencing an outcome variable are identified in the second equation, while taking into account the use of banana hybrids and its determinants.

The decision to use a banana hybrid is defined as a binary outcome (z_j) of an unobserved latent variable (z_j^*). The latent variable underlying this decision is a linear function of its observed ($\mathbf{w}_j$) and unobserved (u_j) determinants, such that:

$$z_j = 1 \text{ if } z_j^* = \mathbf{w}_j'\boldsymbol{\Phi}_j + u_j > 0,$$
$$z_j = 0 \text{ otherwise.}$$

The impact of growing a banana hybrid, expressed as a continuous outcome (y_j), is conditional on a set of independent variables ($\mathbf{x}_j$) and the endogenous binary variable (z_j) indicating whether a hybrid is used:

$$y_j = \mathbf{x}_j'\boldsymbol{\beta}_j + \lambda z_j + \varepsilon_j .$$

These equations are the basis for regression analysis, estimated with maximum likelihood methods. To allow for the identification of the impact of hybrid use on production vulnerability, the decision to grow banana hybrids is estimated with a set of instruments and other covariates hypothesized to influence this choice (all represented by the vector $\mathbf{w}_j$).

Variable Definition

Dependent Variables

The data used for analysis include 260 banana-growing households in Tanzania. Of the sampled households, 20 percent are using one or more banana hybrids in their groves. Following the approach outlined above, the dependent variable for the decision to grow banana hybrids is binary (1 = grow any hybrid; 0 otherwise). Clearly, the dependent variable in the impact equation could be formulated in a number of different ways. The immediate purpose of introducing hybrids was to reduce the vulnerability of households to production losses from banana pests and diseases. In turn, reducing production vulnerability can have important ramifications for consumption and income vulnerability. Reducing production vulnerability could lead to higher or more consistent consumption levels, either directly through meeting subsistence needs or indirectly, through more regular or increased sales and market purchases. Over time, smoother banana production and/or income can accumulate, contributing to changes in wealth status in the community.

In this single-year cross-sectional study, we have measured production vulnerability in terms of expected yield losses from pests and diseases. The dependent variable is defined as the average yield loss expected from biotic pressures considered jointly, including weevils, black Sigatoka, and *Fusarium* wilt. Yield loss is a continuous variable that has been composed from the raw data. First, farmers were asked the incidence of the biotic pressure (in years) they had experienced during their involvement in banana production. Next, enumerators elicited from farmers triangular yield distributions (minimum, maximum, and mode) in the presence and absence of the biotic pressures. Physical differences in the effects of pathogens on different banana cultivars were readily observable to farmers, and photos were provided to assist them in recognition. Expected yield losses were calculated, per biotic constraint and cultivar, from the elicited yield distributions (Hardaker, Huirne, and Anderson 1997). Once composed, the expected yield loss per constraint was averaged across the cultivars grown for each household. Finally, the maximum average expected yield across the three biotic constraints was chosen as a measure of production vulnerability, providing a single indicator of yield loss per household.

Determinants of Hybrid Use

As noted in Chapters 6 and 7, the model of the agricultural household and a broad literature on technology adoption provide the basis for the choice of explanatory variables, comprising individual, household, and physical farm characteristics. Variables are defined and summary statistics provided in Table 9.2.

Among the variables hypothesized to influence the use of banana hybrids are individual and household characteristics, as well as an instrument that aids with the identification of the effect of hybrid use on vulnerability. The gender of the banana production decisionmaker[2] reflects preferences for banana attributes and use. Banana hybrids are multiuse cultivars that can be used for cooking purposes. While women tend to prefer cultivars destined for consumption, men are more likely to use hybrids, which are useful for beer brewing and sale. Years of experience of the banana production decisionmaker in tending for the banana grove, formal education of household members (as an average number of years), as well as the number of extension visits are indicators of acquired human capital in banana-related decisions. Acquired human capital may be positively associated with use of hybrids, providing greater knowledge of the potential benefits associated with them in terms of food provision and income generation. The ratio of dependents (children and the elderly) to active adults reflects the consumption needs of the household. The value of livestock assets and exogenous income are used as proxies for household wealth. In the literature on agricultural innovations, wealthier households are often found to be more likely to grow hybrids, either because of better access to inputs and information or because of greater willingness to bear any risks with new planting material. Among the physical farm characteristics is the size

[2] The banana production decisionmaker is not necessarily the household head, but the person in charge of banana production and management decisions in the household. As in Uganda, this person is often a woman.

Table 9.2 Summary statistics and hypothesized effects of variables used in the analysis

Variable	Description	Mean	Standard deviation
Dependent variables			
Hybrid use	Household grows a hybrid cultivar (yes = 1; no = 0)	0.19	0.39
Yield loss	Average expected yield loss to joint biotic pressures (percent)	4.23	7.31
Explanatory variables			
Gender	Gender of primary production decisionmaker (1 = male)	0.70	0.46
Education	Average aggregate household education level (years of schooling)	6.30	1.98
Experience	Years of experience tending for the banana grove	20.57	13.87
Dependency ratio	Ratio of children and elderly to active adult household members	0.48	0.22
Extension	Number of contacts with extension agents	1.63	4.01
Exogenous income	Income received in previous period (ten thousand Tsh)	20.34	62.69
Livestock assets	Value of total livestock assets (ten thousand Tsh)	20.28	43.92
Farm size	Size of landholding (acres)	1.72	1.43
Elevation	Elevation (stratification variable; 1 = low)	0.85	0.36
Probability BS	Perceived frequency of occurrence of black Sigatoka disease	0.09	0.18
Probability FW	Perceived frequency of occurrence of *Fusarium* wilt disease	0.23	0.23
Probability WE	Perceived frequency of occurrence of weevils	0.34	0.28

of the landholding, another indicator of wealth and the scale of production.

Exposure to banana hybrids and elevation are two stratification variables. Only elevation is included as an instrument in the use equation, allowing for the identification of the treatment effect in the impact equation. The strength of the exposure variable in capturing the effect of use of formally distributed hybrids was compromised by the markedly informal means of transfer of planting material from one farmer to another (this interpretation is supported by descriptive information summarized in Chapter 5). With as many as 20 percent of farmers in nonexposed areas reported to grow banana hybrids, the treatment effect of exposure was dissipated. Geographical location is believed to better explain use behavior, with 96 percent of households growing banana hybrids (that is, 48 of the 50 households using hybrids) residing in low-elevation areas.

Determinants of Production Vulnerability

Among the variables hypothesized to have an effect on production vulnerability are the acquired human capital variables and scale of production (farm size). These characteristics are intended to capture preferences for management of the banana grove and scale of production, which have implications for yield loss, whereas the frequencies of occurrence of the three biotic pressures reflect the direct effects of disease and pest constraints on production vulnerability. Production vulnerability is also hypothesized to depend on the use of banana hybrids. As hybrids are bred for resistance to biotic pressures (in particular to black Sigatoka and weevils), they are expected to reduce the average production vulnerability for households that grow them.

Results

Before the treatment effect model was estimated, the exogeneity of the dichotomous variable for use of banana hybrids was tested in the impact equation. The test was achieved through several steps. First, the use equation was estimated using logit, and the residuals were saved. Then the impact equation was estimated using the actual observations for use of banana hybrids, as well as the saved residuals. No statistical indication of correlation between the errors of the impact and use equations was found (p-value = 0.194 for residuals), supporting the inclusion of use of banana hybrids as an explanatory variable in the impact equation (Woodridge 2001).

The treatment effect model is estimated using maximum likelihood methods. The statistical validity of a simultaneous estimation of both equations is evidenced by the significance of the hazard function (p-value = 0.045 for lambda). The information from one behavioral process (that is, treatment regression) influences the outcomes of another process (that is, the impact equation), similarly to the Mills ratio in a Heckman estimation approach. Estimated coefficients are presented in Table 9.3 separately for the use and impact equations.

The impact equation is the focus of this chapter. Production vulnerability appears to be scale neutral, as farm size has no apparent effect in the impact equation. Larger farmers suffer as much as smaller ones in terms of the percentages of banana production lost to biotic pressures. Nor does accumulated experience managing the banana grove counteract these pressures. The number of years growing bananas has no statistically significant effect on yield losses, illustrating the limited capacity of farmers to combat biotic constraints by applying only established management practices. Aggregate levels of education of the household appear to be positively related to loss, perhaps because it is related to the opportunity cost of labor in terms of managing the grove. Furthermore, higher losses appear to be associated with more extension visits. This finding perhaps is dictated by the need of extension workers to closely monitor the effects of biotic pressures on bananas and to suggest appropriate responses. As expected, more frequent occurrence of black Sigatoka disease and weevils increases the average yield loss. Given the low input practices for banana management, this result supports the dissemination of resistant cultivars in the region for mitigating biotic constraints. The most important result is the statistically significant effect of the use of new banana hybrids on expected yield losses. The magnitude of the effect of using FHIA hybrids is large, demonstrating the benefits to smallholder farmers in Tanzania in terms of reduced production vulnerability.

The probability that a household uses banana hybrids increases with years of schooling of household members, as is commonly found in hybrid adoption studies. More-educated households appear to be more likely to adopt new banana hybrids, which can be attributed to the fact that school curricula in the country address issues of agricultural production management. Wealthier households, in terms of value of owned livestock assets, appear to be more likely to use hybrids. However, rather than a wealth effect, this result likely reveals complementarities between use of hybrids and the provision of fodder (as a by-product of banana production) for animal feed. One key result is the pronounced use of hybrids in low-elevation areas where pest and disease pressures are higher and program intervention would have the highest impact. This result confirms previous work and the findings from this survey (Chapter 9).

Other factors most frequently associated with adoption of high-yielding cultivars are, on average, of no statistical significance for

Table 9.3 Coefficient estimates for the treatment effects model

Variable	Treatment equation (hybrid use)	Impact equation (average expected yield loss)
Gender (1 = male)	0.2069	
	(0.2844)	
Dependency ratio	−0.7227	
	(0.5609)	
Education	0.1764**	0.6709**
	(0.0740)	(0.3018)
Experience	−0.0158	0.0472
	(0.0115)	(0.0407)
Extension	0.0148	0.2158*
	(0.0237)	(0.1259)
Farm size	−0.0059	−0.6427
	(0.1029)	(0.4152)
Elevation (1 = low)	0.9486**	
	(0.4326)	
Probability BS		11.7009***
		(3.5097)
Probability FW		−0.2633
		(2.3900)
Probability WE		8.7904***
		(1.7214)
Exogenous income	0.0001	
	(0.0017)	
Livestock assets	0.0050*	
	(0.0026)	
Hybrid use		−12.2328**
		(5.5963)

Notes: Standard error in parentheses. ***, **, and * indicate statistical significance at the 1, 5, and 10 percent levels, respectively. See Table 9.2 for definitions of the variables.

banana hybrids in Kagera Region. For example, the gender of the production decisionmaker appears to bear no relationship to hybrid adoption when other factors are taken into account. This finding probably reflects the extensive involvement of both men and women in various aspects of banana production and marketing (Chapter 5). Nor does wealth, measured in terms of exogenous income or farm size, affect the adoption decision. Scale and wealth effects in Asia's Green Revolution were often associated with access to water and fertilizer, complementary inputs recommended along with improved wheat or rice seed (Feder 1982). In this case, though labor-intensive management practices had been recommended, the only introduced input was new planting material. Furthermore, the program intervention was a focused, locally organized dissemination effort that distributed planting material free of charge. Consistent with this explanation, extension visits have no perceptible effect on the use of hybrids, because the emphasis of extension agents has been on management practices. Experience with banana production of the primary decisionmaker does not appear to influence the likelihood of using banana hybrids. Neither does the composition of households in terms of the proportion of dependents to active members.

Regression analysis shows the marginal effect of factors on adoption and impact in terms of the average relationship for the sample of households. A comparison of households with a high statistical likelihood of growing banana hybrids and those with a low likelihood of using them, in areas exposed to program interventions, provides different insights into adoption patterns in study villages. Other indications of program impact are revealed through comparing the characteristics of households in the exposed areas, given adoption, and those located in nonexposed areas. These comparisons, made by combining the fitted regression model with descriptive statistics, are presented in the next two sections.

Comparisons of Adopters and Nonadopters in Exposed Areas

First, the fitted model was used to predict hybrid use for the 260 households in the sample. The top and bottom tails of the distribution of predicted values for hybrid use are used to represent the profiles of likely adopters and nonadopters' of the hybrids, considering initially only the subset of 220 households located in the exposed areas. The cutoff point used is 20 percent, with 44 households representing each group, respectively. Mean comparisons are summarized in Table 9.4 for outcomes related to productivity, consumption and sale of bananas, as well as general household-related characteristics.

The comparisons are based on actual values of the outcome variables, with the exception of average yield loss. The treatment model allows us to use predicted values for the estimated outcome variable to control for the effects of treatment on the outcome. Hence, two comparisons are made for the outcome variable: one using the actual values and the other using the predicted values of average yield loss. Although the predicted vulnerability levels are lower than actual vulnerability levels, no significant differences are found between the means for adopters and nonadopters, perhaps because of the wide variation in the sample data. The magnitude of the mean (actual) yield loss is 30 percent higher for nonadopters.

The insights obtained from comparing other outcomes are interesting. The average yield of potential adopters is higher than those less likely to adopt the new hybrids, which is not surprising, given the larger bunch size of hybrids. The total banana production (in number of bunches) is also larger for likely adopters, as their scale of production tends to be larger. No significant differences are found between levels of total and per capita consumption of adopters and nonadopters. However, there is lower consumption per adult in adopting households. These households tend to be larger and have lower dependency ratios, suggesting that consumption is spread across the larger proportion of active members. No differences in consumption per dependent are identified between the two groups.

Market participation behavior of adopters and nonadopters appears to be similar, as is the volume of bunches sold. However, there are significant differences in the distribution of produced bananas in terms of the share the household consumes and sells, although they are similar in the levels they consume and sell. Although no differences in banana income are identified, the farm income and the value of livestock assets of adopting households are higher than those for nonadopters, as adopters also tend to have higher expenditure levels. This result has implications for the dissemination and use of new hybrids in the exposed areas of Kagera Region. Targeting relatively deprived (in terms of income and wealth) households may increase adoption rates of hybrids and could have positive income effects on poor

Table 9.4 Mean comparisons between likely nonadopters and adopters in exposed areas

Variables	Nonadopters (N = 44)	Adopters (N = 44)
Outcomes		
Average yield (kg)	15.45*	18.92*
Actual average yield loss (percent)	6.45	4.50
Predicted average yield loss (percent)	4.33	4.48
Total bunches harvested	171.25	192.25
Number of bunches consumed	95.75*	124.59*
Number of bunches consumed per person	26.66	23.49
Number of bunches consumed per adult	55.51**	34.28**
Number of bunches consumed per dependent	48.06	74.44
Share of bunches consumed (percent)	79.06**	68.06**
Number of bunches sold	71.52	67.73
Share of bunches sold (percent)	20.77**	32.39**
Farm income (ten thousand Tsh)	22.40*	43.08*
Banana income (ten thousand Tsh)	5.86	5.91
Household expenditure (ten thousand Tsh)	23.00*	37.74*
Other characteristics of households		
Exogenous income (ten thousand Tsh)	21.82	39.61
Livestock value (ten thousand Tsh)	8.53***	53.05***
Mean household education levels (years)	4.23***	7.70***
Farm size (acres)	1.38*	1.95*
Household size	5.25*	6.39*
Household dependency ratio	0.55***	0.32***
Number of extension service visits	0.95***	4.04***
Proportion of male-headed households	0.68	0.77
Proportion of households selling bananas	0.61	0.50
Proportion of households buying bananas	0.05	0.07

Note: ***, **, and * indicate statistical significance at the 1, 5, and 10 percent levels, respectively.

households. Adopters are also those households with more acquired human capital, in terms of formal education and contacts with extension agents. More targeted extension visits may be a useful tool for increasing adoption levels of hybrid bananas in the exposed areas. Gender roles are similar between both groups, with hybrid bananas being typically multiuse cultivars often sold, used for banana beer brewing, and, to a limited extent, used for cooking.

Comparisons of Households in Exposed and Nonexposed Areas

Comparisons of average vulnerability across households in exposed and nonexposed areas are presented in Table 9.5. When only those factors that affect the use of banana hybrids are considered, the impact of exposure to the program on expected yield losses from pests and diseases is statistically significant and very large in magnitude (a loss

Table 9.5 Mean vulnerability comparisons between farmers in exposed and nonexposed areas (percent)

Vulnerability outcome	Not exposed ($N = 40$)	Exposed ($N = 220$)
Expected yield loss, considering factors that affect hybrid use and yield loss	2.75	4.21
Expected yield loss, considering only factors that affect hybrid use	8.37***	3.35***

Note: *** indicates statistical significance at the 1 percent level.

of about 3 percent compared to 8 percent). Once hybrid use and other confounding factors affecting yield losses have been taken into account, however, any apparent difference between farmers who have been exposed to the program and those who have not is of no statistical importance. This finding is most likely attributable to the strong role that farmers themselves play in exchanging planting material.

Conclusions

This chapter builds on earlier efforts to document the uptake of newly developed banana hybrids in Tanzania. These hybrids, developed in Honduras, are higher yielding and resistant to the pests and diseases that ravaged banana production in the lakes region in recent decades. Kagera Region, studied here, was particularly affected. In response to this challenge, the governments of Tanzania and Belgium embarked on an ambitious program to introduce disease-free materials. An early paper on the impact of these efforts was produced by KCDP (KCDP 2002; Weedrt 2003).

Our analysis augments the understanding of this process in the following ways. First, the advantage of the treatment model is that it enables us to account for the effects of program exposure on use and to control for variables that affect both hybrid use and program impact. Second, the outcome variables can be used to express any one of numerous aspects of program impact in quantitative terms. Here, we have used expected yield losses from pests and diseases.

Findings demonstrate that the intended impact of reducing yield losses to pests and diseases has been achieved, supporting research efforts aimed at developing resistant planting material and the formal diffusion program. The research findings have important policy implications in favor of ongoing efforts to develop new resistant cultivars bred or engineered to withstand disease pressures. Smoothing production has implications for food security and banana marketing. To sustain these benefits, however, institutional aspects of disseminating new cultivars need to be addressed. In particular, the finding that exposure to formal dissemination programs bore no statistical relationship with uses of hybrids underscores the farmer-based nature of planting-material systems. In part, this finding reflects the mode of reproduction for the crop. Planting-material systems for clonally propagated crops are largely farmer-based. Nevertheless, the finding provides strong indication that for the impacts of introducing new materials to be sustainable, innovative, farmer-based systems should be encouraged and supported.

Improved cultivars that are genetically resistant to pests and disease may not be the best route out of poverty, but as suggested in this study, they do reduce farmers' vulnerability to their surrounding production environment. They also reduce the need for purchased pesticides and fungicides that are costly to farmers, bear health risks for farming communities, and degrade the environment when unregulated.

References

Anandajayasekeram, P., and D. R. Martella. 1996. Conceptual framework to assess the impact of agricultural research. *Journal of Agricultural Economics and Development* (Tanzania) 1 (1): 3–9.

Baijukya, F. P., and E. C. R. Folmer. 1999. Agro-ecological zonation of the Kagera Region. In: *Planning the future: Past, present and future perspectives of land use in the Kagera Region*, ed. E. C. R. Folmer, C. Schouten, and F. P. Baijukya. Bukoba, Tanzania: ARDI-Maruku.

Baker, J. 2000. *Evaluating the impact of development projects on poverty: A handbook for practitioners.* Washington, D.C.: World Bank.

Bosch, Ch. H., A. Lorkeers, M. R. Ndile, and E. Sentozi. 1996. *Diagnostic survey: Constraints to banana productivity in Bukoba and Muleba districts, Kagera Region.* Bukoba, Tanzania, and Amsterdam: ARDI-Maruku and Royal Tropical Institute.

Byabachweze, M. S. R., and A. S. S. Mbwana. 1999. Banana and its improvement efforts at Maruku. In *Planning the future: Past, present and future perspectives of land use in the Kagera Region*, ed. E. C. R. Folmer, C. Schouten, and F. P. Baijukya. Bukoba, Tanzania: ARDI-Maruku.

Byabachweze, M. S. R., P. B. Steenhuijesen, and P. A. Rwezaula. 1997. *Banana seasonality in Karagwe District, a problem of perception.* Field Note 74. ARDI-Maruku, Tanzania.

Doss, C. R., 2003. Understanding farm level technology adoption: Lessons learned from CIMMYT's micro surveys in eastern Africa. International Maize and Wheat Improvement Center (CIMMYT) Economics Working Paper 03-07. Mexico, D.F.: CIMMYT.

Feder, G. 1982. Adoption of interrelated agricultural innovations: Complementarity and the impacts of risk, scale, and credit. *American Journal of Agricultural Economics* 64 (1): 94–101.

Hardaker, J., R. Huirne, and J. Anderson. 1997. *Coping with risk in agriculture.* Wallingford, U.K.: CAB International.

KCDP (Kagera Community Development Programme). 2002. *Annual report.* Bukoba, Tanzania: KCDP.

Lorkeers, A. 1995. Agro-ecological zones of Bukoba District. Applied Soil Fertility Research Project, Soil Section ARDI-Maruku. Manuscript.

Maddala, G. 1983. *Limited-dependent and qualitative variables in econometrics.* Econometric Society Monograph 3. Cambridge: Cambridge University Press.

MAFS (Ministry of Agriculture and Food Security). 2001. *Agricultural sector development strategy. United Republic of Tanzania.* Dar es Salaam: MAFS.

Nkuba, J. M. 2007. Assessing adoption and economic impacts of new banana varieties on livelihoods of farmers Kagera Region, Tanzania. Ph.D. thesis, Sokoine University, Sokoine, Tanzania.

Nkuba, J. M., A. S. S. Mbwana, C. Schouten, and S. Mkulila. 1999. *Testing improved banana varieties in Bukoba District, Kagera Region.* Field Note 96. Bukoba, Tanzania: ARDI-Maruku and Kagera Community Development Programme.

Nkuba, J. M., L. J. Ndege, and S. Mkulila, S. 2002. *On-farm testing and acceptability of improved banana varieties in Bukoba District, Kagera Region (1997–2001).* Field Note 144. Bukoba, Tanzania: ARDI-Maruku.

Ravallion, M. 1994. *Poverty comparisons.* Fundamentals in Pure and Applied Economics 56. Reading, U.K.: Harwood Academic Publishers.

RCO (Regional Commissioner's Office). 2003. *Annual report. Kagera Region.* Bukoba, Tanzania: RCO.

Weedrt, J. 2003. *Adoption of superior banana varieties in the Kagera Region: Accomplishment and constraints.* Bukoba, Tanzania: Kagera Community Development Programme and Economic Development Initiatives Limited.

Woodridge, J. 2001. *Econometric analysis of cross section and panel data.* Cambridge, Mass., U.S.A.: MIT Press.

CHAPTER 10

Assessing the Potential Impact of Selected Technologies on the Banana Industry in Uganda

Robert Kalyebara, Stanley Wood, and Pamela Nahamya Abodi

In this chapter, an economic surplus approach is applied to assess the potential impact of a range of current and emerging technologies. Specialists and experts guided the definition of 6 banana production systems, determined according to productivity potential and the commercial orientation of growers, and 14 technology scenarios that span current best practices for managing bananas, genetic transformation, and conventional breeding. Simulations indicate that current recommended scenarios could generate the highest levels of gross benefits, assuming relatively high rates of adoption. Transgenic varieties resistant to major pests and diseases currently affecting banana production in Uganda appear to generate greater potential benefits than do conventional technologies. In part, this finding reflects expectations that gene insertion is more effective at combating specific pests and diseases. Time lags associated with R&D are also expected to be even longer for conventional banana improvement than for genetic transformation, given the plant's sterility.

Characterizing Banana Production Systems

Six banana production systems were defined for the purpose of this analysis, using the approach taken by specialists in Uganda (Tushemereirwe et al. 2001). This categorization involved the recognition of three classes of geographical area according to their intrinsic biophysical suitability to support banana production (high-, medium-, and low-potential productivity zones), in each of which production is subdivided into two categories according to the production orientation of growers (semicommercial and subsistence). Most producers maintain banana plantations primarily to meet food security and cultural needs, only intermittently entering the market as opportunities and needs arise. There is a much smaller but growing share of farmers whose main focus is to produce for the banana market. Clearly, production and technology use decisions, and response to market opportunities and signals, are distinct for these two groups.

Table 10.1 summarizes the main characteristics of the six systems with respect to the production of cooking bananas (the endemic AAA-EA cultivar group matooke, as well as some exotic FHIA hybrids). Across all areas in Uganda, a total of some 5.5 million tons of cooking bananas are produced from about 450,000 ha of cropland, at a rough national average yield of around 12 tons per ha (UBOS 2002). Beer and sweet cultivars account for an

Table 10.1 Ugandan cooking banana production, by production system, 2000

		High productivity			Medium productivity			Low productivity		
	Total	Total	Semi-commercial	Subsistence	Total	Semi-commercial	Subsistence	Total	Semi-commercial	Subsistence
Production (million tons)	5.545	3.549	2.484	1.065	0.998	0.200	0.798	0.998	0.100	0.898
Production share (percent)	100	64	45	19	18	4	14	18	2	16
Yield (tons/ha)	12.31	21.91	25	17	10	18	9	5.29	11	5
Area harvested (thousand ha)	451	162	99	63	100	11	89	189	9	180
Area share (percent)	100	36	22	14	22	2	20	42	2	40

Sources: Compiled by the authors from the Uganda National Household Survey 1999/2000 (UBOS 2002) and Tushemereirwe et al. (2001).

Notes: The Uganda National Household Survey data on banana production categorizes bananas by use: cooking, beer, and sweet (dessert). This table reports the allocation of the cooking banana production data into the six system categories.

Table 10.2 Estimated yield losses caused by priority banana production constraints in each of the six production systems

Constraint	Typical losses		High productivity		Medium productivity		Low productivity	
	Loss (percent)	Year[a]	Semi-commercial (percent)	Subsistence (percent)	Semi-commercial (percent)	Subsistence (percent)	Semi-commercial (percent)	Subsistence (percent)
Bananaweevil	50–70	Fourth	5	10	15	26	40	60
Nematodes	40–60	Fourth	10	15	14	28	23	51
Black Sigatoka	30–50	Third	6	15	18	28	30	50
Banana bacteria wilt	80–100	First	58	58	58	58	58	90
Low soil fertility	10–70	>Third	37	58	67	83	80	91

Sources: Compiled by the authors from unpublished sources (such as the National Agricultural Research Organization's banana program surveys and reports) and from Rukazambuga, Gold, and Gowen (1998) for weevils, Speijer and Kajumba (1996) for nematodes, and Okaasai and Boa (2004) for bacteria wilt.

Notes: A two-stage process was used to obtain the data. Following an initial review and compilation of available published and unpublished sources, an expert consultation was undertaken to extrapolate available evidence and estimate losses within each production system category.

[a]Year of plantation cycle in which losses typically become significant.

additional 538 thousand tons and 46 thousand tons, respectively, of the total national banana output (UBOS 2002). Although the high-productivity areas account for almost 65 percent of the total production, they account for only about 36 percent of the area planted in bananas. Conversely, the low-productivity areas occupy some 42 percent of the area in bananas, but they provide only some 18 percent of total output. This pattern reflects the disparity in yield levels attained across production systems, ranging from only 5 tons per ha for the low-productivity subsistence systems to around 25 tons per ha among semicommercial producers in high-productivity areas. Low-yield subsistence-oriented production represents 40 percent of the banana producing area, the largest area share among the six groups. Medium- and low-yield semicommercial production each represent only 2 percent of the banana area.

Most of the high-productivity areas lie above 1,200 m.a.s.l., where the climate tends to suppress the virulence of common banana pests and diseases (Speijer et al. 1994; Karugaba and Kimaru 1999; Tushemereirwe et al. 2001). Thus, from a geographic perspective, high-productivity systems are found almost exclusively in Western Region, Uganda, whereas the bulk of the low-productivity systems are found in Central Region (Tushemereirwe et al. 2001, 2003). Low-productivity systems make up some 78 percent, 66 percent, and 21 percent of the production areas in East, Central, and Western regions, respectively. In Central Region, low productivity often reflects overexploitation of soil resources in the past (Gold et al. 1999; Bagamba et al. 2000). In many banana production areas the chronic underreplenishment of soil nutrients appears to have exacerbated vulnerability to pests and diseases (Gold et al. 1999).

Estimates of the losses attributable to the priority production constraints of this study are summarized by production system in Table 10.2. These include weevils (Rukazambuga, Gold, and Gowen 1994, 1998; Gold and Messiaen 2000; Gold et al. 2004), nematodes (Kashaija et al. 1994; Speijer and Kajumba 1996), black Sigatoka (Tushemereirwe et al. 1996), bacteria wilt

(Kangire and Rutherford 2001; Okaasai and Boa 2004), and low soil fertility. Overall, low soil fertility and banana bacteria wilt appear to constitute the greatest threat to banana productivity. Banana bacteria wilt has recently spread alarmingly through much of Central Region with devastating effects (Okaasai and Boa 2004). Just as striking is the assembled evidence that yields on the extensive low-productivity areas occupied by subsistence-oriented growers are reduced by 50 percent or more by a combination of biotic pressures (weevils, nematodes, and black Sigatoka); the same constraints reduce yields by an estimated 5–15 percent in high-productivity areas. This evidence reflects both differences in biophysical conditions and greater use of improved cultural practices in the high-productivity areas (NARO, IITA, and NRI 1994; Tushemereirwe et al. 2003).

Average effects only are summarized in Table 10.2; the dynamic nature of losses and the interaction among constraints are key features of the process underlying low banana yields. With the likely exception of banana bacteria wilt, whose impacts are both short-term and extreme, the magnitude of losses increases over time frames measured in years. Table 10.2 also notes the approximate time after which the cumulative effects of yield losses are generally considered to become economically significant. The pace and intensity at which yield losses increase depend on a range of factors, but their interaction conspires to radically alter the economically viable life of any given mat and, hence, any given plantation. Assessing the economic impact of even a single constraint is complicated by the dynamic interplay of biophysical conditions, management practices, and the coevolution of the banana plant and its constraints. The situation is further complicated by the coexistence of multiple constraints, whose separate effects are difficult to distinguish. To isolate the ameliorating influence of improved management practices on individual constraints is similarly difficult. Thus, although we have attempted to capture the dynamic nature of yield losses and productivity-enhancing interventions, we have not sought to address the interaction of constraints.

Production structures, costs, and returns for different banana production systems are summarized in Table 10.3. In constructing partial farm budgets, data were compiled from a number of baseline studies conducted by the NARO banana research program between 1999 and 2002, including those by Bagamba et al. (2000) for Central Region (primarily low-productivity zones) and Ssenyonga et al. (1999) for Ntungamo and Mbarara districts (high-productivity zones). The data components needed to estimate returns are reported: revenues (annual yield, farm-gate prices), costs (labor, other inputs), and fixed costs (plantation cycle). Though shown in summary form, revenues and costs were calculated over a 30-year time horizon. Plantation life cycles may be as short as 5 years in low-productivity subsistence-oriented areas, and as long as 80 years for well-managed plantations in high-productivity areas. It is noteworthy that in the high-productivity zone, in particular, not only do the more-productive areas utilized for commercial gain have higher yields and longer plantation life cycles, but they also have higher farm-gate prices. This correlation might be attributed to several factors, such as the capacity of growers to provide greater quantities on a more consistent basis, their greater bargaining power, or better bunch quality. Returns to production appear skewed in favor of the more-productive areas and the commercially focused farmers, as reflected in the unit cost of production. In high-productivity areas production costs range from Ush 11–16 per kg of bananas, whereas in all other areas they range from Ush 50–55 per kg. The unit production costs shown in Table 10.3 provide the basis for assessing the potential economic benefits of technology innovations. These innovations are discussed next.

Table 10.3 Annualized banana production costs and returns in each of the six production systems

Budget/cost components	Average	High productivity: Semi-commercial	High productivity: Subsistence	Medium productivity: Semi-commercial	Medium productivity: Subsistence	Low productivity: Semi-commercial	Low productivity: Subsistence
Average plantation cycle (years)	31	80	40	30	20	10	5
Average yield (ton/ha)	12	25	17	18	9	11	5
Farm-gate price (Ush/20-kg bunch)[a]	1,417	2,080	1,680	1,200	1,140	1,200	1,200
Total revenue (Ush/ha)	850,000	2,600,000	1,428,000	1,080,000	513,000	660,000	300,000
Labor cost (Ush/ha)	322,100	266,846	269,456	420,280	364,420	355,300	256,300
Other input cost (Ush/ha)	n.a.[b]	n.a.	n.a.	521,500	126,500	253,000	n.a.
Total cost (Ush/ha)	472,267	266,846	269,456	941,780	490,920	608,300	256,300
Net return (Ush/ha)	624,566	2,333,154	1,158,544	138,220	22,080	51,700	43,700
Unit production cost (Ush/kg)	40	11	16	52	55	55	51

Sources: Derived by authors from National Agricultural Research Organization (NARO) surveys, program documents, and expert knowledge from NARO scientists.

Notes: Budgets were constructed from a complete annual input-and-output analysis over the plantation life cycle (to a maximum of 30 years) for each production system; n.a. indicates not available.

[a]For farm households whose consumption exceeds production, the relevant price is more properly the market purchase price. For the sake of simplicity, however, we have used the farm-gate prices as the basis of all budget-related calculations.

[b]Average not computed because of missing values.

Technology Scenarios

We began by identifying important subcategories of banana production systems as determined by potential productivity and commercial orientation of producers. Based on a characterization scheme used by researchers in Uganda (Tushemereirwe et al. 2001), three banana productivity zones were distinguished (high, medium, and low).[1] Within each zone, semicommercial and subsistence farmers were typified, as shown in Table 10.4, by consulting available data on use practices and technologies. Each of the six subcategories was specified as a separate component of the overall industry supply in the IFPRI-DREAM[2] model as a basis for all technology simulations.

Three broad technology scenario groups were identified through stakeholder consultations: (1) increased adoption of current best practice, (2) conventional crop improvement, and (3) genetic modification. The first group of scenarios is of particular relevance to those engaged in technology transfer and diffusion, such as National Agricultural Advisory Services, the Ugandan publicly funded dissemination agency. Scenarios directed at evaluating strategies for germplasm improvement (for example, conventional versus transgenic) are largely relevant to NARO, those who fund NARO, or those who collaborate with it.[3]

For each scenario, information was compiled to support complete parameterization

[1] The high-productivity zone used in the mesoscale analysis largely corresponds to the >1,200-m.a.s.l. stratum used in the design of the farm-level survey.

[2] IFPRI-DREAM (Dynamic Research Evaluation for Management) is a menu-driven software package for evaluating the economic impacts of agricultural R&D.

[3] In reality the choice of investment strategy is not between conventional and transgenic approaches, but involves both, as explained in Chapter 4. However, scenarios are designed as simple (often polar) examples to demonstrate the sensitivity of returns to actions, situations, or assumptions.

Table 10.4 Predominant management practices, by production system category

Zone	Farm type	Predominant management practice
High productivity	Subsistence	Crop sanitation by majority; light mulching by a few
	Semicommercial	Crop sanitation + mulching + manure/compost
Medium productivity	Subsistence	Crop sanitation by majority; mulching with residues of banana and other crops
	Semicommercial	Crop sanitation + mulching + manure/compost; a few apply fertilizer (less than 5 percent)
Low productivity	Subsistence	Most plots managed with no sanitation, mulch, or manure/compost
	Semicommercial	Crop sanitation + mulching + manure/compost

Sources: Tushemereirwe et al. (2001); NAADS (2003).

Table 10.5 Targeting of technologies relevant to industry-scale scenarios

Constraint	Principal target genomic group	Target environment	Technology option
Wilts	All	Highland/lowland / high soil fertility	Transgenic/conventional + improved practices
Weevils	AAA-EA	Highland/lowland / low soil fertility	Transgenic/conventional + improved practices
Nematodes	AAA, ABB	Lowland / high soil fertility	Transgenic or conventional
Black Sigatoka	AAA-EA, ABB	Lowland / low soil fertility	Transgenic
Soil fertility decline	All	All	Transgenic or conventional

Note: See Chapter 3 for definitions of the genomic groups.

of the DREAM model, including banana plantation production cost data (with and without constraints), estimates of the likely time lags for and probability of research success, and the potential efficacy of new technologies under different production conditions. Technology data were elicited from scientists and other experts using a survey instrument designed for the purpose, and on-farm costs were compiled mostly from NARO published and unpublished sources. Technology options to address specific constraints were targeted by environment and genomic group, as indicated schematically in Table 10.5.

A total of 14 technology scenarios were implemented in the mesoscale analysis (Table 10.6). In each of the three groups, the four major biotic constraints (weevils, nematodes, black Sigatoka, and bacteria wilt) were considered (12 scenarios). Two additional scenarios were defined in the current best practice group: increased adoption of soil-enhancing technologies and practices (such as the use of compost and mulching), and the adoption of multipurpose FHIA hybrids (see Chapters 4, 5, and 9, and Appendixes A–C for further information about diffusion and adoption).

Key research parameters are also summarized by scenario in Table 10.6. These include the probability of research success, the likely time until research successes are achieved, and the share of yield losses that

Table 10.6 Definition of technology scenarios and selected research parameters

Technology scenario	Probability of research success[a] (percent)	R&D time lag[b] (years)	Potential yield loss avoided[c] (percent)
Scenario Group 1: Increased adoption of current best practice			
1. Improved agronomic and IPM practices for management of weevils	n.a.	n.a.	40
2. Improved agronomic and IPM practices for management of nematodes	n.a.	n.a.	30
3. Improved agronomic and IPM practices for management of black Sigatoka	n.a.	n.a.	30
4. Improved agronomic and IPM practices for management of bacteria wilt	n.a.	n.a.	30
5. Improved nutrient management: inorganic and organic fertilizers, mulch	n.a.	n.a.	80
6. Adoption of introduced multipurpose hybrids (such as FHIA17)	n.a.	n.a.	80
Scenario Group 2: Conventional crop improvement			
7. Improved resistance to weevils	63	15	38
8. Improved resistance to nematodes	48	15	29
9. Improved resistance to black Sigatoka	78	15	21
10. Improved resistance to bacteria wilt	18	15	13
Scenario Group 3: Genetic modification			
11. Genotype with resistance to weevils	78	12	63
12. Genotype with resistance to nematodes	53	12	63
13. Genotype with resistance to black Sigatoka	68	12	54
14. Genotype with resistance to bacteria wilt	68	12	71

Source: Extracted from expert consultation survey results database.

Notes: These estimates were used in setting upper bounds on losses at the end of the plantation life cycle. IPM denotes integrated pest management.

[a]The most likely probability of achieving research success for the specified technology (technologies for Scenario Group I are already available by definition).

[b]The most likely lag time in years before research would be successful. In the case of biotechnologies, this value includes time for regulatory testing and approval processes.

[c]Indicative of value of potential yield losses that could be avoided if the specific technology were successfully developed and adopted. More specific assessments of avoided yield losses (expressed as decreases in the unit cost of production) were made for each technology scenario for each of the six banana production systems over a system-specific plantation life cycle.

the technology innovation might be able to mitigate. As depicted in Figure 2.3 of Chapter 2, the innovation mitigates but does not eliminate the negative impacts of biotic constraints on banana yields. In addition, because technologies and practices already exist for the current best practice (Group 1) scenario, the probability of research success and R&D time lags are irrelevant.

Experts considered that improved resistance to black Sigatoka through conventional breeding and improved resistance to weevils through transgenic approaches are the most likely scientific outcomes (78 percent probability of success). They considered the development of bacteria wilt resistance through conventional means (18 percent) to be least likely. Banana improvement through conventional approaches is expected to take longer (15 years) than through genetic transformation (12 years), because of the challenges described in Chapter 4. Thus, time lags for genetic modification scenarios are shorter, even allowing

Table 10.7 Time to peak adoption and peak adoption level, by scenario (years/percent)

	High productivity		Medium productivity		Low productivity	
Technology scenario[a]	Semi-commercial	Subsistence	Semi-commercial	Subsistence	Semi-commercial	Subsistence
1	2/85	3/63	2/71	3/51	4/56	7/33
2	2/85	3/63	2/71	3/51	4/56	7/33
3	2/85	3/63	2/71	3/51	4/56	7/33
4	2/85	3/63	2/71	3/51	4/56	7/33
5	2/48	3/39	2/36	3/28	4/26	7/17
6	7/23	11/13	6/34	10/23	6/38	8/27
7	7/50	9/36	3/53	5/34	3/65	5/48
8	5/48	8/30	5/49	7/34	4/58	7/35
9	7/35	9/25	6/54	7/34	3/90	4/70
10	3/70	4/50	3/70	4/50	3/60	4/40
11	9/30	13/16	5/38	7/24	5/44	7/31
12	6/34	10/15	6/36	8/20	6/45	8/28
13	9/20	13/10	8/36	9/21	4/69	5/54
14	5/63	7/55	4/55	5/45	4/55	3/45

Source: Median values extracted from expert consultation survey undertaken by this study.
Notes: Table entries show two values: time lag to peak adoption (years)/peak level of adoption (percent) following the successful development and release of a new technology. Current best practice (scenarios 1–6) are already adopted to varying degrees in different production systems, and the maximum adoption levels reported here are the expected cumulative maximum including current levels of adoption. Expected time lags and adoption levels are predicated on maintained investment in dissemination services.
[a]See Table 10.6 for definitions of the technology scenarios.

for delays caused by regulatory procedures associated with the testing, screening, and registration of transgenic materials.

Extensive information about the characteristics and potential impacts of technologies, by production systems and over time, was collected through the expert survey.[4] Table 10.7 reports only two of these parameters, time to peak level of adoption from technology release and the expected peak level of adoption, by scenario. In the case of current best practice, this time lag is relative to 2005. Projected adoption lag times range from 2 years for current best practice use by semicommercial farmers to 10–13 years for improved germplasm use by subsistence farmers. Projected adoption levels range from 85 percent of semicommercial farmers in high-productivity areas using current best practice to only 10–20 percent of subsistence farmers using any of the improved technologies.

The elicitation process also generated information on constraint type, production system, constraint severity, scale of mitigation possibilities, and plantation life cycles. These data enable the construction of profiles of potential yield losses avoided (unit cost reduction). Yield loss profiles are then combined with the lag time and adoption

[4] The survey instrument will be included as an annex in the mesoscale research report.

Figure 10.1 Yield loss avoided by adoption of improved practices and/or technologies for different production system plantation life cycles: Scenario Group 1 (current best practice), increased adoption of improved agronomic and integrated pest management practices for the management of weevils

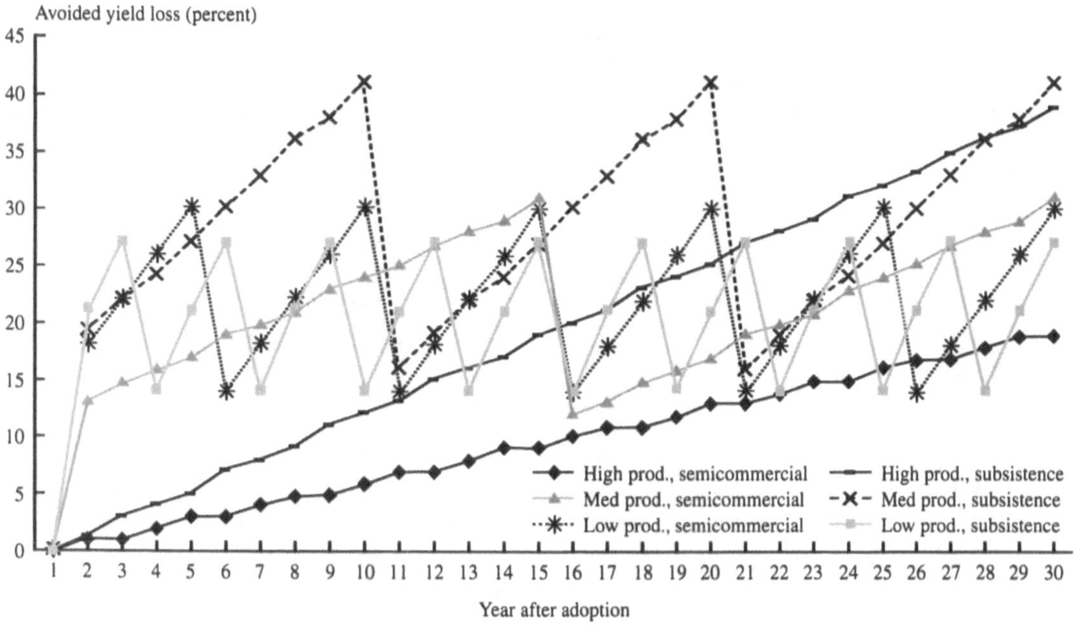

information to conduct the simulations. Figure 10.1 illustrates yield loss profiles for one of the scenarios (CBP for management of weevils). The pattern of yield loss is driven by the plantation cycle associated with each production system (Table 10.3). In the case of a semicommercial production system in a high-productivity area (the lowest curve in the figure), there is a slowly accumulating yield loss within a plantation cycle that is 30 years or more in length. At the other extreme, plantation cycles for the low-productivity subsistence system might be 3–5 years because of rapid build-up of yield losses. Such profiles were generated and applied to each of the 14 scenarios.

Economic Evaluation

As described in Chapter 3, the economic assessment was conducted using the DREAM economic surplus model. The 14 scenarios were individually parameterized, each comprising 7 "regions": the 6 production systems and a single consumption region. A closed economy formulation was used, and a 30-year simulation (2000 to 2030) was run for each scenario in annual time steps. Market prices were determined endogenously in the model by clearing demand and supply each year. Banana demand was projected to grow over the 30-year period, in line with expectations for population growth, income growth, and the income elasticity of banana. For all simulations a supply elasticity of 0.5 and a demand elasticity of –1.0 were used.[5]

For each scenario the model parameterization included a base set of production and market conditions for each region, characterization of the technology and its adoption time lags and profiles, and costs of technol-

[5] The DREAM database is available from the authors on request.

Table 10.8 Present value of gross benefits by banana technology scenario, Uganda (thousand US$)

	Producer benefits								
	High productivity		Medium productivity		Low productivity				
Technology scenario[a]	Semi-commercial	Subsistence	Semi-commercial	Subsistence	Semi-commercial	Subsistence	Total producer benefits	Total consumer benefits	Total benefits
4	578,410	145,123	40,118	87,337	13,578	22,828	887,394	494,026	1,381,420
5	371,697	114,302	47,446	106,419	17,683	18,809	676,356	371,840	1,048,196
14	237,903	65,795	13,916	18,433	4,096	15,914	356,057	351,765	707,822
6	211,415	8,677	31,256	38,810	10,350	40,584	341,092	338,911	680,003
1	56,410	63,365	24,478	103,878	9,989	–26,313	231,807	123,427	355,234
2	66,916	70,085	19,434	65,071	8,821	–12,774	217,553	118,386	335,939
3	21,600	28,739	7,621	72,245	9,659	–5,354	134,510	69,600	204,110
12	104,334	182	8,155	2,905	3,667	8,657	127,900	126,460	254,360
11	80,196	482	11,502	12,008	4,723	18,501	127,412	126,451	253,863
13	10,545	–8,638	9,778	10,612	10,190	62,289	94,776	91,704	186,480
7	51,701	7,857	5,839	4,042	2,543	10,664	82,646	82,164	164,810
10	37,617	7,695	2,504	2,180	432	–401	50,027	49,772	99,799
8	35,755	3,297	2,566	1,280	1,388	4,643	48,929	48,706	97,635
9	7,754	–974	4,450	4,940	3,741	23,011	42,922	42,255	85,177

Notes: Table entries are present values of gross benefits computed over a fixed 30-year period, even though different technologies have different lag times before their potential benefits may be realized (for example, current best practices are available for adoption now, whereas conventional breeding approaches may require up to 15 years before technologies become available for adoption). A discount rate of 10 percent per year was used in all cases.

[a]See Table 10.6 for definitions of the technology scenarios.

ogy generation and dissemination (costs of current best practices were not included—these were treated as sunk costs). Some of the data elements used in this parameterization are reported in the tables above. The analysis proceeds as an annual simulation in which the future situation with regard to the production, consumption, and price of banana is computed with and without the introduction of a new technology. The difference between the economic impacts of these two simulations on a yearly basis determines the gross annual research benefits attributable to the intervention. Table 10.8 presents the results of this analysis in terms of the present value of gross benefits.

Simulation results are shown in Table 10.8, ordered in terms of decreasing gross benefits to banana growers. Producer benefits range from an estimated US$887 million for current best practice (bacterial wilt) to an estimated US$43 million for conventional crop improvement (black Sigatoka), a benefits ratio of 20:1. Findings suggest that the highest benefits will be attained through improved adoption of current best practice, followed by genetic transformation. Of the conventional crop improvement options, resistance to weevils is the most favorable.

One obvious explanation for the superiority of current best practice is that a technology available now that can provide significant yield savings is likely to be more attractive than solutions available 10–15 years in the future, because of the time value of money. Furthermore, the economic returns will be even larger than the ratios in gross benefits, because there will be no offsetting costs associated with technologies already developed.

The only other group of technology that ranks in the top half by size of producer benefits is genetic modification to mitigate the effects of banana bacteria wilt. This technology has a shorter expected lag time than conventional breeding, and the scale of the damage it could help mitigate is large.

Purely conventional approaches to crop improvement appear to offer the lowest potential benefits (although they still might be economically attractive, providing a high rate of return to research investment once costs are taken into account). This finding does not, of course, signify that conventional breeding is no longer needed. Genetic modification relies fundamentally on many conventional crop improvement activities, meaning that the transgenic group of scenarios might better be labeled as "genetic modification and conventional." Rather, the results indicate that exclusively following conventional approaches would be the least desirable improvement strategy, primarily because biotechnology appears to offer some unique solutions to the complex breeding issues related to sterility of bananas (Chapter 4).

Two other aspects of the results warrant further attention. The first is that consumer benefits are also quite large, sometimes as large as producer benefits. This finding arises from the downward pressure on prices exerted by improved productivity, relative to the situation without the new technology. Indeed, though prices may rise throughout the period covered by our simulations, prices will rise less rapidly than is the case if technological change had not taken place. The total benefit of the technological change is the sum of both producer and consumer benefits.

A second aspect of these results is that negative producer benefits are occasionally generated by technology scenarios. Thus, relative to the benefits they would have received without technical change, a group of producers receives fewer benefits with technical change—a decrease in welfare for that group. This situation can arise for a number of reasons. The most common is that other producer groups are adopting innovations earlier or at a faster rate and deriving greater benefits from lower unit production costs. Another reason is that technologies might be biased in the sense that they deliver greater impacts in specific production systems, and producers in other systems cannot take full advantage of them.

Conclusions

We have assessed the potential economic benefits of a range of technology options that are available to R&D policymakers and managers in Uganda. Our results suggest that "doing the easy things first" (better use of knowledge already gained) may have the highest payoffs and presumably might present the fewest implementation challenges.

Pursuing this option alone is not enough, however. More needs to be done, and done relatively soon, to raise the productivity of the banana sector, given its importance in the diet of Ugandans and the large amount of agricultural land currently allocated to relatively unproductive banana systems (and that could, by implication, be reallocated to more economically productive uses). At the same time, it is already a challenge to maintain existing productivity levels. Pests and diseases evolve, and new challenges are faced in consolidating past gains. This realization alone must drive the urgent need to maintain and preferably expand the effort to mitigate the biotic pressures that cause large economic losses.

Our results further suggest that accelerating the emphasis on a combined transgenic-conventional improvement strategy has several advantages, not least of which is time. If transgenic solutions can be derived more quickly through such a mixed approach, that strategy merits support.

In terms of specific biotic constraints, it is less easy to pick winners among the technologies currently under development. Mitigation of the effects of bacteria wilt appears to provide very large payoffs, especially if that disease should continue its current damaging trajectory. Improved soil fertility, both in its own right and as a means of improving resistance to biotic constraints and ensuring long-term production sustainability, appears to exhibit high potential payoffs (second only to mitigation of bacteria wilt). Increased adoption of the multipurpose FHIA hybrids might have high payoffs from the perspective of an increased stock of more resistant materials, but appears unlikely to happen without changes in consumer preferences (Chapter 5). Prioritizing among resistance to weevils, nematodes, and black Sigatoka is particularly problematic, given the confounding effects and/or simultaneous occurrence of these constraints, as noted in the farm-level analysis and previous research (Chapter 6). This difficulty is one evident reason why NARO has favored tackling multiple constraints simultaneously. Were the analysis able to take account of the interaction effects of constraints, the economic results might be quite different.

There remain no shortage of methodological and data issues surrounding the analysis performed here, including the consideration of research investment and diffusion costs. The development of a complete cost series was not possible within the time frame of this research. Nevertheless, the possibility of performing such improved evaluations has been very greatly enhanced by the databases, methods, and analysis applied to the current analysis by Ugandan scientists and their collaborators.

References

Bagamba, F., J. Ssenyonga, W. K. Tushemereirwe, E. Katungi, C. S. Gold, and A. Katwijukye. 2000. Characterization of banana production systems in central Uganda. Research report. National Banana Research Programme, Kawanda, Uganda. Manuscript.

Gold, C. S., and S. Messiaen. 2000. *The banana weevil* Cosmopolites sordidus. *Musa* Pest Fact Sheet 4. Montpellier, France: International Network for the Improvement of Banana and Plantains.

Gold, C. S., E. B. Karamura, A. Kiggundu, F. Bagamba, and A. M. K. Abera. 1999. Monograph on geographic shifts in highland cooking banana (*Musa*, group AAA-EA) production in Uganda. *African Crop Science Journal* 7 (3): 223–298.

Gold, C. S., G. H. Kagezi, G. Night, and P. E. Ragama. 2004. The effects of banana weevil, *Cosmopolites sordidus,* damage on highland banana growth, yield and stand duration in Uganda. *Annals of Applied Biology* 145 (3): 263–269.

Kangire, A., and M. Rutherford. 2001. *Wilt-like disorder of bananas in Uganda. Musa* Disease Fact Sheet 10. Montpellier, France: International Network for the Improvement of Banana and Plantains.

Karugaba, A., with G. Kimaru. 1999. *Banana production in Uganda: An essential food and cash crop.* Regional Land Management Unit Technical Handbook Series 18. Uppsala, Sweden: Regional Land Management Unit and Swedish International Development Cooperation Agency.

Kashaija, I. N., P. R. Speijer, C. S. Gold, and S. R. Gowen. 1994. Occurrence, distribution and abundance of plant parasitic nematodes on bananas in Uganda: Preliminary results of a diagnostic survey. *African Crop Science Journal* 1 (2): 99–104.

National Agricultural Advisory Services (NAADS). 2003. Profitability analysis of agricultural enterprises in Uganda. Report prepared by M. Wamamiala and P. Wathum, Independent Consulting Group, Uganda.

NARO, IITA, and NRI (National Agricultural Research Organization, International Institute for Tropical Agriculture, and Natural Resources Institute). 1994. Working document. Diagnostic survey on key constraints of banana production in Uganda. Preliminary data on socio-economics, soils and agronomy, pests, diseases and plant growth, germplasm, post harvest. Ugandan National Banana Research Programme, IITA, and NRI, Kawanda, Uganda. Manuscript.

Okaasai, O., and E. Boa. 2004. Distribution of banana bacterial wilt. Ministry of Agriculture, Animal Industry, and Fisheries, Republic of Uganda. Available at: http://www.agriculture.go.ug/bananawilt.php.

Rukazambuga, N. D. T. M., C. S. Gold, and S. R. Gowen. 1994. Banana weevil-host plant (Musa AAA-EA) interactions in East African Highland banana systems. In *African Crop Science Conference proceedings,* ed. E. Adipala, M. A. Bekunda, J. S. Tenywa, M. W. Ogenga-Latigo, and J. O. Mugah, June 14–18, 1993, Kampala, Uganda. Kampala: African Crop Science Society.

———. 1998. Yield loss in East African highland banana (*Musa* spp., AAA-EA group) caused by the banana weevil, *Cosmopolites sordidus* Germar. *Crop Protection* 17 (7): 581–589.

Speijer, P. R., and C. Kajumba. 1996. Yield loss from plant parasitic nematodes in East African highland banana (*Musa* spp., AAA). *MusAfrica* 10: 26.

Speijer, P. R., C. S. Gold, E. B. Karamura, and J. N. Kashaija. 1994. Banana weevil and nematode distribution patterns in highland banana systems in Uganda: Preliminary results from a diagnostic survey. In *African Crop Science Conference proceedings,* ed. E. Adipala, M. A. Bekunda, J. S. Tenywa, M. W. Ogenga-Latigo, and J. O. Mugah, June 14–18, 1993, Kampala, Uganda. Kampala: African Crop Science Society.

Ssenyonga, J. W., F. Bagamba, C. S. Gold, W. K. Tushemereirwe, E. B. Karamura, and E. Katungi. 1999. Understanding current banana production management in south-western Uganda. In *Mobilizing IPM for sustainable banana production in Africa,* ed. E. A. Frison, E. B. Karamura, and R. A. Sikora. Workshop proceedings, November 23–28, 1998, Pretoria, South Africa. Montpellier, France: International Network for the Improvement of Banana and Plantains.

Tushemereirwe, W. K., M. Holderness, C. S. Gold, E. B. Karamura, and M. Deadman. 1996. Distribution and importance of the leafspot complex on highland bananas in Uganda. *MusAfrica* 10: 25.

Tushemereirwe, W. K., D. A. Karamura, H. Sali, D. Bwamiki, I. Kashaija, C. Nankinga, F. Bagamba, A. Kangire, and R. Sebuliba. 2001. Bananas (*Musa* spp.). In *Agriculture in Uganda.* Vol. 11, *Crops,* ed. J. Mukiibi. Kampala, Uganda, and Wageningen, The Netherlands: Fountain Publishers, Technical Centre for Agricultual and Rural Cooperation, and National Agricultural Research Organization.

Tushemereirwe, W. K., I. N. Kashaija, W. Tinzaara, C. Nankinga, and S. New, eds. 2003. A *guide to successful banana production in Uganda. A banana production manual.* Kampala, Uganda: National Banana Research Programme, Kawanda Agricultural Research Institute.

UBOS (Uganda Bureau of Statistics). 2002. *Ugandan National Household Survey 1999/2000: Report on the Crop Survey module.* Entebbe: UBOS.

Part IV. Conclusions

CHAPTER 11

Conclusions and Implications for Research Policy

Melinda Smale, Wilberforce Kateera Tushemereirwe, Pamela Nahamya Abodi, Fredrick Bagamba, Mgenzi Said Ramadhan Byabachwezi, Svetlana Edmeades, Robert Kalyebara, Enid Katungi, Enoch Mutebi Kikulwe, Jackson M. Nkuba, and Stanley Wood

The research compiled in this report documents the uptake of newly released banana hybrids and recommended banana-management practices on semi-subsistence smallholder farms in the Lake Victoria region of Uganda and Tanzania. A baseline of 800 household farms was drawn with a random sample stratified by village exposure to new banana varieties and elevation, a factor that is related to soil fertility and the incidence of pests and diseases. The baseline is used to characterize banana varieties, farmers, and villages statistically. Building on previous taxonomic work and the baseline data, a complete taxonomy of distinct banana varieties grown in the region was developed.

Econometric models were applied to the baseline data to test hypotheses related to four aspects of banana productivity and production constraints, with implications for banana improvement. A stochastic production frontier was estimated to assess banana production efficiency, considering soil fertility and labor constraints. Probit, ordinary least squares, and Heckman models were used to explain the use and extent of use of soil fertility management practices, considering the role of social capital, in addition to farm, household, and market characteristics. Farmer demand for planting material of transgenic banana varieties with resistance to pests and disease, currently under development by the research program, was predicted. A treatment model was then applied to evaluate the farm-level impact of adopting hybrid bananas. The potential impact of farmer adoption of transgenic varieties on the banana industry, and the social distribution of economic benefits from adoption, were then estimated in the economic surplus framework.

Similarities and Contrasts between Uganda and Tanzania

Several economic and biophysical features of banana production are both common to the Lake Victoria region of Uganda and Tanzania and set them apart from banana production areas in other regions of the world. The same genomic groups, including the endemic highland banana and two nonendemic groups, exotics and recently bred hybrids, are grown in this region. A great diversity of varieties and clone sets, distinguished by agromorphological descriptors, is found at the farm, village, and regional scales. Varieties are also classified locally by their use in cooking, brewing of local beer, roasting, or sweet desserts (snacks).

A common set of biophysical pressures constrain banana productivity in the region. In both Tanzania and Uganda, drastically declining yields were reported from the 1970s to the early 1990s, caused by soil fertility problems, weevils, nematodes, *Fusarium* wilt, and black Sigatoka. Since then (and not studied here), bacteria wilt has beset banana plantations in the region. Other economic factors pose challenges for raising banana productivity and encouraging the product commercialization that will lead to increased rural income. Continuous propagation in banana groves implies high labor demands—in the face of rising opportunity costs for labor—to manage soils and mats effectively. Perennial production means that bananas often serve a key role in meeting food consumption or cash needs, particularly for poor families, but commercialization is inhibited by the bulk and perishable nature of the product. Many bananas are sold at the farm gate, with high and variable market margins. Although there is potential for product diversification through processing, a negligible percentage of production is now processed. The prospects for exports are modest, given the preferences of world consumers for dessert bananas, and the competition.

Against this common backdrop, each country has pursued its own strategy for resolving the challenges. The Ugandan National Banana Research Programme (NBRP) has pursued a three-pronged approach to combating biotic constraints to banana production. In addition to assembly and evaluation of local and foreign germplasm, including hybrids from FHIA and IITA, the program has undertaken breeding for resistance by crossing and genetic transformation, combined with tissue culture propagation of clean planting material. Elite highland bananas have also been propagated and distributed to farmers. The best-yielding highland bananas have proved sterile and hence not amenable to conventional breeding. As a consequence, genetic transformation is a major option for attaining host-plant resistance. Related research is under way on a range of varieties that represent the diversity among genomes and clone sets in the region, in order to retain the end-use attributes appreciated by consumers. The third element of NBRP's approach is the development and recommendation of banana management practices, consisting of both natural resource (soil and water) and sanitation techniques.

Like the national research program in Uganda, the program in Tanzania has brought in a large number and wide variety of materials for evaluation under local conditions. In the context of the KCDP project, the governments of Tanzania and Belgium then diffused relatively large amounts of new planting material in Kagera Region, including exotics and hybrids.

Farmer participation in evaluation and testing, as well as their central role in supplying planting material, is a salient feature of the traditional production system and is reflected in the strategies of both NARO and ARDI. Tanzanian researchers have documented the role that farmers have played historically in locating sources of clean planting material—often over great distances—over the past half-century recalled by farmers in personal interviews. Various farmer-based systems have served as a means of efficient dissemination of materials, albeit on small scales in specific locations in Uganda. Not only the material, but also the information about management practices appears to diffuse primarily through farmer-based channels, although roughly half of farmers in Uganda rely on formal sources in addition to other farmers. Of formal sources, government extension figures most prominently.

Linked by language, culture, and history, banana growers in Tanzania and Uganda are separated by a national border and live in different administrative and political jurisdictions. In both the Ugandan and Tanzanian subdomains, roughly one-quarter of household heads are women. Both men and women participate in cultivar choice, final production decisions, sales, and beer making, although there seems to be a more pronounced gender division of labor in some tasks in Tanzania.

Participation in banana markets is patchy, underscoring the semisubsistence nature of banana production for many households in the region. The rate of commercial sales appears much higher in the highlands of Tanzania than in those of Uganda, and in both countries, the highlands regions are much more commercially oriented than the lowlands.

In Uganda, statistical tests confirm that expected yield losses from black Sigatoka differ by elevation, with the geographical focus of the disease in the lowlands. Consistent with researchers' observations, cooking quality and yield (bunch size) are the characteristics that growers consider most important. Endemic varieties are considered superior for cooking, and hybrids are rated higher for bunch size and resistance to weevils, though perceived differences for disease are not statistically significant in Uganda. Hybrids are rated more highly across all attributes than endemic types in Tanzania. As signaled by the Tanzanian authors in this report, disease pressures were so great in some areas of Kagera Region that farmers lost their endemic varieties entirely, leaving no point of reference.

Differences in dissemination strategies, biotic pressures, farmer perceptions, and the geographical scale of the area studied probably contribute to the stark differences in adoption rates between Uganda and Tanzania. In Uganda, the percentage of farmers using hybrids is negligible when considered across the full expanse of banana-growing areas, but within targeted villages, use rates are high. In Tanzania, the 20 varieties most frequently grown in Kagera Region include one of the FHIA hybrids, and roughly one in five farmers has planted banana hybrids. Nearly all farmers grow exotic farmers' cultivars. The rate of use of recommended natural resource and mat management practices is much higher in Uganda than the rate of use of improved materials, although the labor intensity of these practices means that typically, only a portion of the average plantation is tended using the practice.

The focal role of farmers in planting material systems, emphasized in the chapters on banana improvement strategies in both countries, is confirmed by the baseline data. In Uganda, there are large numbers of exchanges of planting material of all types, but more so for nonendemic bananas, including hybrids bought and sold for cash. Most planting materials are exchanged without money, in part because initial dissemination by KCDP was free of charge. The fact that Ugandan farmers traversed an average of 15 km for new planting material, compared to less than 3 km in Tanzania, indicates that high fixed costs of transaction may also be impeding adoption in Uganda.

The application of econometric models supports several general conclusions related to adoption of existing and emerging technologies. First, findings demonstrated the potentially pro-poor application of transgenic cooking varieties and showed that the choice of host cultivar for trait insertion is likely to have social consequences. Simulations illustrate the extent to which supporting public investments in education, market infrastructure, and extension will augment farmer demand for new planting material. The demand for planting material of the cooking varieties that are potential host varieties for gene insertion varies according to the characteristics of households, farms, and markets and the attributes of varieties. In particular, farmers also demand material with lower expected yield losses to black Sigatoka and weevils—traits targeted for genetic transformation.

Second, the evidence confirms that adoption of FHIA hybrids reduces the vulnerability of Tanzanian households to yield losses from pests and diseases. For farmers with few nonfarm sources of income, reducing production risk can smooth both consumption and income from local sales. The baseline data indicate that direct exposure to new varieties through formal programs is no longer a critical determinant of their use in Tanzania, evidence once again that farmer-to-farmer exchange of planting materials is crucial. The econometric model confirms this result. Heterogeneity among households in terms of social, demographic, or wealth characteristics appears to have less importance in use decisions than does the incidence of pests and disease—which motivates farmers' search for new materials in this production system.

Third, it is evident that social capital plays a significant role in the use of current best practice for managing soil fertility in banana production in Uganda. Most of the villages in the banana producing areas foster active social organization, with numerous associations and high rates of membership, although membership in economic associations is more exclusive. Village social characteristics will likely have especially important implications for planting material systems for EAHBs, because transfers of planting material and related information are based on farmers themselves rather than on a formal seed industry.

The productivity of cooking bananas and the factors that determine it depend on the economic, climate, and soil characteristics of regions, and as hypothesized, it is highest in the southwest, least in Central Region, and intermediate in Masaka. There is some evidence that the efficiency of banana production can be improved, particularly in Central Region. Labor productivity is higher in Central Region, where most agronomic practices (such as crop sanitation) are carried out minimally. Soil pH and the application of manure as recommended have positive and significant effects on productivity, especially in the southwest and Central Region of Uganda. The critical importance of labor is evident in findings regarding the age of the household head and family size. Road infrastructure and production credit will enhance efficiency.

Findings illustrate substantial inefficiencies in the production of cooking bananas, especially in central Uganda. Education improves technical efficiency in the southwest but not in central Uganda. Market access (using distance to paved roads as a proxy) improves efficiency for the sample farmers in Masaka and southwest but not in the Central Region, where opportunity costs of labor are higher. Household size is positively related to efficiency but is only significant for central Uganda. Banana production appears to be more efficient when managed by men than by women, probably due to underlying differentials in access to resources. Access to credit increases efficiency in the Central Region, whereas rent and remittances improve efficiency in Masaka and the southwest.

Policies to improve production efficiency include investments in education, extension, and infrastructure (roads), and improving access to production credit for commercially oriented producers in the southwest and Masaka areas. Clearly, policies will need to be tailored to the local conditions, as demonstrated by the different production and efficiency profiles of the three regions analyzed here.

The analysis of banana production efficiency reveals the presence of surplus labor in the highlands, whereas access to farm labor is a constraint in the lowlands. Wages are higher for casual labor in the lowlands, where there is a more developed market for unskilled labor. Investment in technology development could raise banana productivity in the highlands. In the lowlands, both technology development and extension education have positive effects on productivity. Investment in human capital (especially women's education) and enabling better access to input and credit markets could improve banana production; investments in the paved road network would improve the comparative advantage of the region in banana production. Intercropping bananas could be interfering with the implementation of certain practices (such as mulching, timely weeding, and crop sanitation).

Simulations of the gross economic benefits that could be generated by a set of technology options indicate that more widespread adoption of current best practice is likely to generate the greatest investment payoffs. Yet the authors argue that the longer term strategy endorsed by NARO, which combines conventional and transgenic approaches to mitigate the biotic pressures that cause major economic losses, is essential for sustaining banana production systems. A good research policy should provide for both short- and long-term goals, as is currently the case at NARO. With regard to the potential for genetic transformation of banana, we cannot interpret the findings in terms of projected dollar benefits alone. Discussions with researchers clearly indicated that given the sterility of the endemic cooking banana, future crop improvement will be limited unless transgenic tools are applied. Although the initial costs of establishing a biotechnology capacity are very high in terms of infrastructure and human capital development, as the capacity to plan and undertake biotechnology-based research is strengthened, there will be multiplier and spillover effects in other research areas.

Policy Implications

The results of this study have implications for the R&D policies that currently influence the banana industry in East Africa. These can be grouped into five categories:

1. *Developing improved banana genotypes.* Findings confirm the vulnerability of the endemic banana genotypes to pests and/or diseases and the need for a research policy to commit long-term investments in development of resistant banana genotypes. The choice of host cultivar, and use group, will have social consequences in terms of the farmers most likely to use and benefit from the technology. In general, however, greater social costs are likely to be caused by delays in banana improvement in terms of benefits foregone. Time lags in research and adoption have often been shown to be the single most important parameter determining the social payoff to investment.
2. *Enhancing demand and supply of improved germplasm.* A policy supporting investments in agricultural education, extension, marketing infrastructure, and access to good roads will enhance demand and supply of improved banana varieties and in turn raise banana productivity and efficiency. We predict a demand for pest- and disease-resistant material, given the evidence that farmers value these traits, but demand will be much greater if supporting investments in education, extension, and market infrastructure are also made. Successful gene insertion in a cultivar that is already familiar to farmers will make the genetically modified cultivar more attractive to such farmers and to those who resemble them in terms of social, economic, and farm characteristics, but other investments will alter the likelihood of use and benefits distribution. There is an increase in farmer demand for planting material when market prices improve, even though they remain largely subsistence oriented, indicating the potential for increased commercial production in areas of comparative advantage.
3. *Optimizing dissemination mechanisms.* Findings support the current policy of the Government of Uganda (through the National Agricultural Advisory Services), which emphasizes farmer associations and human capital development as pillars of technology and knowledge dissemination. Human, but also social, capital and other active dissemination mechanisms are important for perceptions, and perceptions are important for the recommended practices that are labor intensive and

for which the "technology" is generated on the farm. Again, farmers are price responsive, undertaking more production when the output price relative to the input price is high. The high rates of dissemination of improved soil fertility practices are promising, despite the labor these practices demand. Labor constraints are less of a problem in the highlands, the more commercially oriented production zone with labor surpluses. Farmer- and association-based mechanisms appear to be a crucial factor in disseminating both planting material and technologies.

4. *Scaling up the adoption of genotypes.* There is need for a demand-driven strategy for scaling up farmer use of approved banana varieties. There is widespread adoption of FHIA hybrids in Tanzania, given the great disease pressures, extensive dissemination efforts, and the historical practices of farmers in actively seeking pest- and disease-free planting material. Adoption definitely shows an impact on vulnerability to disease losses. Further analysis is needed over time to see whether diffusion is sustained, benefits are sustained, and income effects are observable. The examples of farmer-to-farmer exchanges described for Uganda, though limited in their impact in terms of numbers of farmers and communities, warrant closer examination as models for more structured, though more decentralized, diffusion mechanisms. We recommend a farmer-based and socially based network design, with farmer-supplied planting material, possibly scaling up from some of the experiences in Uganda. The strategy of providing large quantities of materials free of charge (a "push" strategy) is not sustainable.
5. *Determining R&D strategies for the highlands and lowlands.* Strategies for improving productivity in high-elevation areas will necessarily be different from those targeting the low-elevation areas. The highlands require development and promotion of best cultural practice and marketing improvements, whereas lowlands require development and promotion of pest- and/or disease-resistant endemic varieties together with best cultural practice aimed at reviving productivity. To support the success of these efforts, major investments need to be made in dissemination. Overall, we recommend targeting more than one trait, but not too many simultaneously, because such a strategy could delay transformation, curbing the benefits to be earned by Ugandan society.

Technical change is a continuous, multidimensional process. Social science can contribute to the formulation of agricultural research policy during technology development by identifying the social, economic, and institutional impediments that must be addressed to ensure that promising new crop technologies will be adopted by farmers. Once these technologies are in the course of adoption, social science can provide insights into mechanisms that will support adoption rates and enhance the positive impact of technical change on the livelihoods of rural people.

The National Banana Research Programme of NARO and IITA, with the support of the INIBAP, have embarked on an ambitious breeding program that employs a range of biotechnologies (embryo rescue and somatic embryogenesis) as well as crossing, propagation, and crop and soil management practices, to address the most debilitating diseases and pests. Some of the improved technology explored in this report already exists and some is emerging from these efforts. The baseline developed here, which represents a single, statistically based "snapshot" in a gradual process of technical change, has been used to predict the ex ante impact of transgenic cultivars. In addition, it has been used to assess the ex post impact of conventional hybrids, investigate the role of social capital in the adoption of practices to enhance soil fertility, and assess the efficiency of banana production. Later, this same baseline will be available to assess the ex post impacts of emerging technologies. The sector analysis provides an overview of the potential distribution of benefits across the range of technologies currently promoted or under development.

APPENDIX A

Banana Taxonomy for Uganda

Svetlana Edmeades and Deborah Karamura

A combination of the survey data collected for the current and previous research (Karamura 1998; Karamura and Pickersgill 1999) was employed to construct this taxonomy. Based on previous taxonomic research, a subset of key banana descriptors was preselected for inclusion in the sample survey. Farmers surveyed named the varieties planted in their banana groves during the seasons of the survey and described them in terms of each characteristic (bunch size, bunch position, finger compactness, size of fingers, maturation period, pseudostem color, and space between the hands of the bunch). Responses were then cross-checked with existing taxonomic knowledge. Household shares and cultivar shares were then estimated from the survey data.

Table A.1 Banana taxonomy, Uganda

	Cultivar name	Synonyms	Genomic group	Common use	Household share (percent)	Cultivar share (percent)
1	Atwalira nyina	Nasaba	1	1	0.77	0.03
2	Bogoya	Musiideke	22	3	41.01	2.49
	(Gros Michel)	Musindije	(AAA)			
		Eringot				
3	Bogoya Omumyufu	Ekijungu (Red Bogoya)	22	3	2.71	0.13
		Epiakol	(AAA)			
4	Bikumpu	Mukubakonde	1	1	4.06	0.51
		Nfuunya bikonde				
5	Butobe	Entobe	1	1	10.83	1.84
		Kafunze				
		Bujonjo				
6	Ekakot	—	1	1	0.19	0.00
7	Ekuron	Okuron	1	1	1.16	0.06
8	Embururu Embiire	—	1	2	0.58	0.01
9	Engongo	Rwamugongo	1	1	4.26	0.38
10	Enkara	Entundu(a)	1	4	5.42	0.42
		Ntuundhu				
		Rwasha				

(continued)

Table A.1 (continued)

	Cultivar name	Synonyms	Genomic group	Common use	Household share (percent)	Cultivar share (percent)
11	Enkyonkyo	—	1	1	0.77	0.02
12	Ensenyuka	—	1	2	0.77	0.05
13	Enshenyi	—	1	1	3.87	4.01
14	Enshenyi Embiire	—	1	2	2.71	0.45
15	Ensika	Engumba	1	2	1.55	0.04
16	Ensowe	Nsowe	1	2	2.90	0.16
17	Enzirabahima	Kalyankoko	1	4	7.74	0.77
		Kibalawo				
		Enjuumba				
18	Entukura	Engabani	1	2	3.29	0.20
19	Enyabotembe	—	1	2	2.32	0.15
20	Enyaruyonga	Naluyonga	1	1	4.06	0.17
21	Enzinga	Rugata	1	1	0.39	0.01
		Nalwetinga				
22	Gonja	Wette	22	5	14.31	0.66
			(AAB)			
23	Kabula	Enyamaizi	1	2	5.80	0.52
		Namaji				
24	Kaburuka	—	1	1	2.13	0.05
25	Kalyankoko	—	1	1	0.77	0.03
26	Kamenyaggali	Kahendagari	1	1	0.58	0.15
27	Kapusente	—	1	4	0.97	0.02
28	Kasese	—	1	1	0.19	0.00
29	Katalibwambuzi	—	1	2	0.58	0.01
30	Katwalo	Entanzinduka	1	1	2.90	0.23
		Enzoga				
		Kashenga				
31	Kawanda (FHIA)		23	4	4.45	1.01
		FHIA01	(AAAB)			
		FHIA03	(AABB)			
		FHIA17	(AAAA)			
		FHIA23	(AAAA)			
32	Kayinja	—	22	2	14.31	4.04
			(ABB)			
33	Kazirikwe	—	1	1	0.19	0.00
34	Keitabunyonyi	—	1	1	0.19	0.01
35	Kibuzi	Enshansha	1	1	32.50	6.38
36	Kidhozi	—	22	4	9.48	0.79
			(ABB)			
37	Kininira	—	1	1	1.55	0.07
38	Kisubi	Egero-gero	22	2	28.43	3.99
		Kanyamwenge	(AB)			

Table A.1 (continued)

	Cultivar name	Synonyms	Genomic group	Common use	Household share (percent)	Cultivar share (percent)
39	Kivuvu	Boki-boki	22	4	14.31	1.17
		Ruhumbo	(ABB)			
		Ekalimon				
40	Kiyovu	Kisansa	1	1	11.61	2.10
		Namayovu				
		Nakayovu				
		Kayovu				
		Mayovu				
41	Km5	—	23	4	0.77	0.03
			(AAA)			
42	Lwaddungu	Ntika	1	1	5.03	0.99
		Ntwiika				
43	Lyewudhika	Nalwewunzika	1	1	0.39	0.02
44	Majaga	—	22	5	0.39	0.06
			(AAB)			
45	Makunku	Bukumo	1	1	2.51	0.05
		Bukunko				
46	Malira Nakasabira	—	1	1	0.77	0.02
47	Malira Nalugiri	Soolabasezaala	1	1	1.93	0.13
48	Malira Nalwela	—	1	1	2.51	0.10
49	Malira Omwirugavu	—	1	1	0.19	0.04
50	Malira Rufuta	—	1	1	0.39	0.04
51	Malira Tatakange	Mukale Sitakange	1	1	0.19	0.01
52	Manjaya	—	22	5	0.19	0.00
			(AAB)			
53	Mbarara	Rwambarara	1	1	0.58	0.22
54	Mbidde	Embiire				
		Ibide	1	2	20.12	4.79
55	Mbwazirume	—	1	1	37.33	4.92
56	Mudwale	Mpologoma				
		Mbale				
		Batule				
		Basimirayo	1	1	5.80	1.64
57	Mugeso	—	1	1	0.39	0.01
58	Mukadde alikisa	—	1	1	0.97	0.37
59	Mukazimugumba	—	1	1	0.58	0.01
60	Musa	—	22			
			(ABB)	2	15.86	4.07
61	Musakala	Nsaagala				
		Namasagala				
		Enshakara				
		Luwata				
		Mayogi	1	1	32.88	4.30

(continued)

Table A.1 (continued)

	Cultivar name	Synonyms	Genomic group	Common use	Household share (percent)	Cultivar share (percent)
62	Muvubo	Saga saga				
		Muzhuba				
		Muzubwe				
		Mujuba	1	1	22.44	1.70
63	Nabusa	Enyeru				
		Enyarweru	1	1	22.63	6.04
64	Nakabinyi	Kakono	1	1	3.09	0.14
65	Nakabululu	Mbululu				
		Embururu				
		Butende				
		Enyigit	1	1	43.52	6.39
66	Nakamali	Malira Omutono				
		Malira Nakangu	1	1	6.96	1.48
67	Nakasabira	—	1	1	1.55	0.05
68	Nakawere	Karinga				
		Kasenene				
		Musenene	1	1	2.13	0.15
69	Nakinyika	Kifuba				
		Kafuba				
		Enzuma	1	1	19.92	2.53
70	Nakitembe	Entaragaza				
		Enshembashembe				
		Malira Omunene				
		Malira	1	1	57.83	9.18
71	Nakyetengu	Kitetengwa				
		Kitika	1	1	5.22	0.49
72	Nakyewogoola	—	1	1	0.19	0.00
73	Nalugolima	Nyarugoroma	1	1	0.39	0.02
74	Nalukira	Enyarukira	1	2	1.35	0.30
75	Nalwesaanya	—	1	2	0.19	0.01
76	Nalyewurula	—	1	1	1.16	0.04
77	Namadhi	Nalusi	1	1	1.35	0.01
78	Namadhugudha	Namunwe				
		Nyeko-ger	1	1	7.54	0.66
79	Namafura	—	1	1	0.19	0.00
80	Namaliga	Namalevu				
		Kyanakyandiga				
		Kiriga				
		Rwakashita	1	1	4.06	0.28
81	Nambi	—	1	1	0.97	0.08

Table A.1 (continued)

	Cultivar name	Synonyms	Genomic group	Common use	Household share (percent)	Cultivar share (percent)
82	Namwezi	Serunjogi				
		Ngalodabha Mbedha	1	1	5.03	0.36
83	Nandigobe	Enjagata				
		Nyarwanda				
		Ntinti	1	1	18.96	1.55
84	Ndyabalangira	Enzirabushera				
		Mulangira				
		Muzirankanja				
		Muziranyama				
		Katetema	1	1	25.73	3.89
85	Nfuuka	Nyakahangazi	1	1	6.00	1.07
86	Njoya	—	1	1	2.13	0.83
87	Nkago	Ikago	1	1	3.09	0.10
88	Nsabaana	—	1	1	0.19	0.00
89	Ntujo	—	1	1	0.19	0.00
90	Okiteng	—	1	1	0.19	0.01
91	Oringoi	—	1	1	0.19	0.01
92	Shombobureku	—	1	1	0.39	0.01
93	Siira	—	1	1	12.38	0.95
94	Ssalalugazi	—	1	1	0.39	0.01
95	Sukali Ndiizi	Kabaragara				
		Osukari				
		Epusit	22			
		Kapere	(AAB)	3	60.74	6.71

Notes: Genomic groups are defined as 1 = endemic (AAA-EA), 2 = nonendemic, 22 = exotic, and 23 = hybrid. Common uses are 1 = cooking, 2 = beer brewing, 3 = dessert, 4 = multiuse, and 5 = roasting. Household share is the proportion of households that grow this cultivar out of 517 households in the sample. Cultivar share is the proportion of mats planted to this cultivar out of 52,321 mats in the sample. — indicates a cultivar without a synonym.

References

Karamura, D. A. 1998. Numerical taxonomic studies of the East African highland bananas (Musa AAA-East Africa) in Uganda. Ph.D. thesis, University of Reading, Reading, England.

Karamura, D. A., and B. Pickersgill. 1999. A classification of the clones of East African highland bananas (*Musa*) found in Uganda. *Plant Genetic Resources Newsletter* 119: 1–6.

APPENDIX B

Banana Taxonomy for Tanzania

Svetlana Edmeades and Deborah Karamura

As in the case of Uganda, a subset of key banana descriptors was preselected to include in the sample survey based on previous taxonomic research. Farmers surveyed named the cultivars planted in their banana groves during the survey season and described them in terms of each characteristic (bunch size, bunch position, finger compactness, size of fingers, maturation period, pseudostem color, and space between the hands of the bunch). Responses were then cross-checked with existing taxonomic knowledge. Exotics and hybrids are classified as different subsets of nonendemic bananas. Survey data were then used to calculate household and cultivar shares.

Table B.1 Banana taxonomy, Tanzania

	Cultivar name	Synonyms	Genomic group	Common use	Household share (percent)	Cultivar share (percent)
1	Akanana	Akanana/mti	22	3	6.61	3.31
		Bunana	(AB)			
		Kanana				
2	Basiga	Basiga-Ekyobazaire	1	2	0.04	0.01
3	Bita3	—	23	4	0.04	0.01
4	Bushwela	—	1	1	0.04	0.01
5	Cardaba	—	22	4	0.11	0.02
			(ABB)			
6	Chiliga	Kiliga	1	1	0.11	0.02
7	Ebihumbu	—	1	1	0.07	0.02
8	Echibetenga	—	1	1	0.04	0.01
9	Ekikonjwa	(E)Kisharuka	22	4	2.69	1.18
		(E)Kivuvu	(ABB)			
		Engalambi				
		Ekigalambi				
		Kikonjwa				
		Kategombwa				
		Ekitegombwa				
		Kikojozi				
		Njologomi				

Table B.1 (continued)

	Cultivar name	Synonyms	Genomic group	Common use	Household share (percent)	Cultivar share (percent)
10	Embilabile	(E)Mbilabile(a)	1	1	0.90	1.26
11	Embile	—	1	2	1.87	1.06
12	Embululu	Nakabululu Enakabululu Kabulu	1	1	0.26	0.09
13	Embwailuma	Mwanaalifa Mwanakufa Mbwazirume Mbwail(r)uma Mbwairuma Mwankule	1	1	3.36	2.43
14	Emfumbo	—	1	1	0.22	0.31
15	Empigi	—	1	1	0.26	0.04
16	Empindwi	Mpindu	1	1	0.22	0.25
17	Enakasese	—	1	2	0.07	0.05
18	Enchoncho	En(y)abusa End(e)aisya Enyambo	1	1	4.86	7.75
19	Endekula	—	1	1	0.04	0.001
20	Endemela	—	1	2	0.07	0.02
21	Enduluma	—	—[a]	2	0.04	0.004
22	Endumuza	—	—[a]	1	0.41	0.88
23	Engagala	Endibwa Endibwabali Endibwabarungi Ndiwa Engagara	1	1	1.38	2.12
24	Enganda	—	1	1	0.07	0.50
25	Engumba	—	1	2	0.82	0.39
26	Enjoga	Enjogo Enjooga	1	1	0.34	0.11
27	Enjoki	Kajoki	1	1	0.15	0.09
28	Enjubo	Mujubu Njumbo	1	1	3.40	1.74
29	Enjujuzi	Enjunjuzi	1	1	0.49	0.11
30	Enjuma	(E)Njumba Kanuma	1	1	0.11	0.04
31	Enjuta	Kajunta	1	1	0.04	0.02
32	Enkila	Enkira Enchila	1	2	0.30	0.16
33	Enkonjwa	Enkonjwa/kya Nkonjwa	22 (AAB)	5	3.32	0.71

(continued)

Table B.1 (continued)

	Cultivar name	Synonyms	Genomic group	Common use	Household share (percent)	Cultivar share (percent)
34	Enkundakundi	Enkundi Enkungu Enkungwe Makungwe	1	2	0.90	1.44
35	Enkunku	—	1	2	0.07	0.01
36	Ensaka	Enshaka	1	2	0.15	0.08
37	Ensenene	—	1	1	0.15	0.03
38	Enshakala	(E)Nshakara E(n)kande	1	1	8.44	11.09
39	Enshalila	—	1	4	0.04	0.02
40	Enshansha	Enshushu	1	1	4.07	4.53
41	Enshanshambile	—	1	2	0.19	0.26
42	Ensikila	Entalio Endalimo	1	1	2.50	2.81
43	Ensowe	Nshowe	1	2	0.19	0.05
44	Entaba	—	1	1	0.04	0.01
45	Entabula	Enigwa	1	2	0.07	0.03
46	Entalagaza	Nakitembe	1	1	2.24	3.70
47	Entama	Entamba	1	1	0.19	0.02
48	Entandala	—	1	1	0.78	0.52
49	Entebateba	Entembatembe Entembe	1	4	0.41	0.09
50	Entente	—	1	1	1.83	5.23
51	Entobe	—	1	1	6.16	7.31
52	Entundu	Ntundu	1	2	3.03	7.80
53	Enyabutembe	—	1	1	0.04	0.11
54	Enyaisheshe	—	1	1	0.04	0.27
55	Enyakasa	Enyakasha	1	1	0.26	0.07
56	Enyalihonga	Enyaruyonga Nziranyonga	1	1	0.22	0.19
57	Enyamaizi	—	1	2	0.41	0.01
58	Enyamawa	—	1	1	0.11	0.04
59	Enyamwonyo	—	1	1	0.07	0.08
60	Enyeisige	Enyesiige Enysiige	1	1	0.22	0.004
61	Enyesiige/mbile	—	1	2	0.04	0.74
62	Enyitabunyonyi	Eyitabunyonyi Kaitabunyonyi	1	1	0.93	0.05
63	Enyonza	—	1	1	0.07	6.27
64	Enyoya	—	1	1	5.38	0.21

Table B.1 (continued)

	Cultivar name	Synonyms	Genomic group	Common use	Household share (percent)	Cultivar share (percent)
65	Enzilabahima	Entuku(la) Kinyantuku	1	1	0.41	0.25
66	Enzimola	—	1	1	0.04	0.01
67	Enzinga	—	1	1	0.07	0.12
68	Enzirabushera	—	1	1	0.26	0.16
69	FHIA	various FHIA01 (Goldi) FHIA02 FHIA17 FHIA21 FHIA23 FHIA25	23 (AAAB) (AAAB) (AAAA) (AAAB) (AAAA) (AAAA)	various FHIA01=4 FHIA02=3 FHIA17=3 FHIA21=5 FHIA23=3 FHIA25=2	2.84	0.64
70	Kainja	(Kisubi in Uganda)	22 (AB)	2	3.66	5.17
71	Kamoshi	—	—[a]	2	0.07	0.01
72	Kauni	Mahuni	1	1	0.04	0.004
73	Kibuzi	—	1	1	0.37	1.42
74	Kibwakyentobe	—	1	4	0.04	0.07
75	Kijoge	Musa (E)Njoge	22 (AAA)	4	7.02	7.49
76	Kiliomunyungu	—	22 (AAA)	4	0.04	0.01
77	Kintu	Jintu Kintubalizibwa	1	4	0.45	0.08
78	Kinunu	—	1	1	0.15	0.02
79	Kisubi	(Kainja in Uganda) Banana Enshubi	22 (ABB)	2	3.06	1.99
80	Km5	Ya(n)gambi	22 (AAA)	4	1.76	0.66
81	Kyanabalya	—	1	1	0.04	n.a.
82	Kyayaya	Kyayaya/n	22 (AAB)	5	0.71	0.21
83	Kyomulutwa	—	1	1	0.11	0.05
84	Kyoya	—	1	1	0.11	0.15
85	Lwekilo	—	1	1	0.07	0.005
86	Mashule	Nshule	1	1	0.04	0.001
87	Mbuzi	—	1	1	0.07	0.02
88	Mkazi Mugumba	Mkazimgumba	1	1	0.15	0.06
89	Mpilwa	—	1	1	0.04	0.02
90	Mpima	—	1	1	0.04	0.01

(continued)

Table B.1 (continued)

	Cultivar name	Synonyms	Genomic group	Common use	Household share (percent)	Cultivar share (percent)
91	Mshale	—	22 (AA)	5	0.34	0.05
92	Mtwishe	—	22 (AAA)	3	4.26	3.33
93	Muhima	—	1	1	0.15	0.02
94	Mwanamwana	—	1	1	0.04	0.03
95	Nteyichungu	—	1	1	0.04	0.01
96	Nyabwehogola	—	1	1	0.04	0.01
97	Nyarujoju	—	1	1	0.04	0.01
98	Pelipita	Peripita	22 (ABB)	4	0.45	0.06
99	Rose	Rozi	22 (AA)	3	0.15	0.02
100	Rwabuganga	—	1	2	0.22	0.04
101	Rwakagoye	—	1	1	0.04	0.01
102	Shilingi	—	22	4	0.56	0.12

Notes: Genomic groups are defined as 1 = endemic (AAA-EA), 2 = nonendemic, 22 = exotic, and 23 = hybrid. Common uses are 1 = cooking, 2 = beer brewing, 3 = dessert, 4 = multiuse, and 5 = roasting. Household share is the proportion of households that grow this cultivar out of 260 households in the sample. Cultivar share is the proportion of mats planted to this cultivar out of 80,317 mats in the sample. — indicates a cultivar without a synonym; n.a. indicates not available.

[a]The genomic group for this cultivar was not identified.

APPENDIX C

Use of Improved Banana Varieties by Village in Uganda

Enoch Mutebi Kikulwe

Improved (exotic) banana varieties were found in 9 of the 27 survey villages of Uganda (3 in eastern Uganda and 6 in central Uganda. Elite endemic varieties were found in 14 villages.

Table C.1 Share of farmers growing improved (exotic and elite) banana varieties (percent)

Village	FHIA01	FHIA03	FHIA17	FHIA21	FHIA 23	Km 5	All exotics[a]	Elites[b]
Kasamba	—	—	—	—	—	—	—	5.00
Bubemba	5.00	—	—	—	—	—	5.00	25.00
Bugaddu	—	5.00	—	—	—	—	5.00	—
Lumundi	—	—	—	—	—	—	—	10.00
Buwaate	—	—	—	—	—	—	—	5.00
Birongo	—	—	—	—	—	—	—	10.00
Nakapinyi	20.00	—	—	—	—	—	20.00	—
Kitegoba	—	—	5.00	—	—	—	5.00	10.00
Nangulwe	—	—	—	—	—	—	—	5.00
Nakulabye	5.00	—	—	—	—	5.00	10.00	10.00
Busige	30.00	65.00	55.00	—	—	15.00	65.00	10.00
Akoboi	—	—	—	—	—	—	—	5.00
Kyembazzi	5.00	—	—	—	—	—	5.00	5.00
Kalembe B (430)	5.00	—	—	—	5.00	—	10.00	—
Mpumudde (432)	—	—	—	—	—	—	—	5.00
Lusolo (434)	—	—	—	—	—	—	—	35.00
Oriyoi A (437)	5.00	5.00	—	15.00	—	10.00	15.00	10.00

Note: — indicates none grown by farmers.

[a] Includes data for all banana plots on all farms surveyed in the village.

[b] Data are based on two major banana plots and all the elite varieties were aggregated (Mpologoma, Namaliga, and Kisansa) per banana plot schedule.

APPENDIX D

Details of Sample Survey Design

Melinda Smale, Pamela Nahamya Abodi, Ingrid Rhinehart, Svetlana Edmeades, Mgenzi Said Ramadhan Byabachwezi, Charles A. Eledu, Kephas Nowakunda, and Stanley Wood

As explained in Part II, Research Context, the introduction of improved varieties to farmers in selected banana-growing areas is one of the approaches among those currently recommended by national and international organizations and NGOs to counter banana pests, diseases, and declining soil fertility. This appendix details the sampling frame employed to document an adoption baseline for improved varieties and management practices, and to diagnose impediments to present and future use.

Population Domain

The domain is purposely selected to cover areas specializing in banana production, including those with currently declining, increasing, and intermediate levels of production. These correspond roughly to the central and southwest geographical zones in Uganda, and Kagera Region of Tanzania. Municipalities (urban areas) were excluded from the domain as unrepresentative. Southern districts of Kagera Region were also omitted, because they presented costly logistical challenges. Districts in Kagera include Bukoba and Karagwe.

A geo-referenced map of East Africa indicating principal banana growing areas was provided by INIBAP to delineate the domain. The EAHB (*Musa* AAA-EA) is the dominant genomic group in the region, featuring mainly in the highland landscapes of Rwanda, Burundi, eastern Democratic Republic of Congo (DRC), Uganda, western Tanzania, and Kenya (Chandler 1995). The group includes two major use classes (cooking banana, or *matooke*, and beer banana, or *mbidde*), a number of clone sets and a large number of cultivars (Karamura and Karamura 1994; Appendixes A and B).

Stratum Definition

Stratification improves sampling efficiency to the extent that it captures major differences in parameters influencing the variability of the characteristics to be measured, and if it results in groups that are relatively homogeneous with respect to those characteristics. The key characteristics estimated in this research are adoption of current materials and predicted adoption of pest- and disease-resistance materials. Because there are trade-offs in precision as the number of stratifying variables increases with a fixed sample size, we confined the variables to two.

The first is environmental. A critical parameter for the adoption of varieties with improved resistance to pests and disease is the yield advantage attained relative to other banana varieties.

In addition to farmers' management practices, which are part of farmer decision-making and thus cannot serve as a control variable, relative yield advantages depend on the disease pressure and productivity potential of the growing environment. In consultation with INIBAP, IITA, NARO, and ARDI scientists, elevation was selected to represent the numerous correlated factors that affect the incidence and severity of most pests and diseases of bananas in the Lake Victoria region (Speijer et al. 1994). Elevation is also related to soil quality, climate, and the surrounding vegetation in these environments (Tushemereirwe et al. 2001). Low elevation was defined as below 1,200 m.a.s.l. and high elevation was defined as above this level.

The second stratifying variable is institutional: previous exposure to new banana varieties (exposed or not exposed). Areas of exposure were defined as LC3s or wards where researchers or extension or other program agents had introduced improved planting material (banana suckers) in at least one community. Areas with no exposure are those where no organized program designed to diffuse improved planting material has been conducted, according to personal consultations with NARO and ARDI, and KCDP records. Areas included in exposed strata represent the factual and those included in the nonexposed strata represent the counterfactual in predicting impacts of improved banana varieties.

Four strata were delineated (i = elevation, j = exposure): (1) low elevation, with exposure ($i = 1, j = 1$); (2) low elevation, without exposure ($i = 1, j = 0$); (3) high elevation, with exposure ($i = 2, j = 1$); and (4) high elevation, without exposure ($i = 2, j = 0$). Other factors known to contribute to variation in adoption probabilities (such as market access and agroecological zone) were not used as criteria for stratification, because large differences in variation were not readily observable at the level of the administrative units sampled and there were no a priori grounds for selection of a cutoff point. Instead, market access was measured continuously in the household survey instruments.

Geo-referenced data about banana production systems, a digital elevation model, maps of administrative units, and information concerning previous diffusion of banana planting material were used to disaggregate the domain into the four strata. The domain and four strata were then mapped onto the administrative level of ward in Tanzania and LC3 in Uganda. Wards and LC3s were designated as high or low elevation, based on a simple majority proportion of the unit being above or below 1,200 m.a.s.l. ARDI knowledge of any exposure in the area (even if only one instance) classified the ward or LC3 as exposed, and complete lack of exposure classified the ward or LC3 as nonexposed.

It is important to note that the administrative level of ward in Tanzania is a relatively large area. Although the size of these units is not necessarily optimal for our sampling purposes, it is the lowest administrative level for which we were able to obtain accurate digitized geographic information. This constraint results in a much larger area to be aggregated for purposes of stratification. Therefore, the possibility of a selected village in a ward to be of high elevation or exposed when the ward is classified as low elevation or nonexposed (or vice versa) is higher in the Tanzania portion of the survey area than in the Uganda portion.

Allocation of Primary Sampling Units to Strata

An efficient allocation of sample to strata in formal sampling schemes is one that minimizes variation within the stratum and maximizes variation between the strata, in turn minimizing overall sampling error (Hansen, Hurwitz, and Madow 1953). When the variances in population parameters are known, the sample can be allocated opti-

Table D.1 Sampling fractions for primary sampling units in the survey domain

Population of primary sampling unit				Sample of primary sampling units			
	Elevation				Elevation		
Stratum	Low	High	Total	Stratum	Low	High	Total
Exposed	49	5	54	Exposed	17	3	20
Row (percent)	(91)	(9)	(100)	s.f.	0.367	0.40	0.37
Not exposed	155	49	204	Not exposed	15	5	20
Row (percent)	(76)	(24)	(100)	s.f	0.097	0.10	0.098
Total	204	54	258	Total	32	8	40

Note: s.f. is the primary sampling unit sampling fraction (n_{ij}/N_{ij}), where i =elevation (1 or 2) and j = exposure (1 or 0).

mally within and among PSUs by choosing their number and the number of households per PSU to minimize the survey (sampling and nonsampling) error, given a fixed budget (De Groote 1996). In this case, the variances in the multiple population parameters of research interest were unknown.

The minimum sample size for conducting hypothesis tests on variables measured at the community level (such as social capital and some physical capital) is 20 each in exposed and nonexposed areas (corresponding to a student's t-test). Though a larger sample of communities would have been preferred for statistical precision, the cost of conducting the research in more than 40 communities scattered across the domain exceeded the budget. The total number of PSUs was therefore fixed at 40, with half distributed in the exposed areas and the other half in nonexposed. The 40 primary sampling units were then allocated between the two elevation levels and two countries proportionate to the probability of selection.

Selection of Primary Sampling Units

PSUs were drawn using systematic random sampling from a list frame with a random start. The sampling fractions for the primary sampling units among the four strata in the domain are shown in Table D.1.

Of the 40 PSUs, 3 in Uganda were purposely selected (Ntungamo, Bamunanika, and Kisekka) to complement soil research conducted by NARO (Uganda), in the context of the thesis research by Bagamba. For the thesis research conducted by Katungi (see Chapter 1), 15 PSUs were selected from this sampling frame, including the 3 purposely selected sites, along with 5 others randomly selected from a domain defined by a project titled "Reviving Bananas in the Central Region," implemented by NARO, NBRP.

PSU sampling fractions (*s.f.*) vary by stratum and are defined as the ratio of stratum-specific sample size (n_{ij}) and stratum-specific population size (N_{ij}), expressed as (n_{ij}/N_{ij}). The final sample of Uganda consists of 27 PSUs, of which 18 are located in nonexposed and 9 in exposed areas. In Tanzania, out of the 13 selected PSUs, 11 are located in exposed and 2 in nonexposed areas. Table D.2 summarizes the distribution of PSUs across the strata and country.

The spatial representation of the primary sampling units is shown in Figure 2.1, with sampled sites highlighted. Figure 2.2 shows only the sites surveyed.

Table D.2 Number of survey sites (primary sampling units), by elevation and diffusion status, Uganda and Tanzania

	Stratum				
	Low elevation		High elevation		
Country	Not exposed	Exposed	Not exposed	Exposed	Total
Uganda	14	8	4	1	27
Tanzania	1	9	1	2	13
Total	15	17	5	3	40

Selection of Secondary Sampling Units

The SSU both in Uganda and Tanzania is a village. In Uganda, in each LC3, there are several parishes (LC2s), and each parish consists of several villages (LC1s). One SSU was selected per PSU. The probability of selection (or sampling fraction) of a SSU varies by PSU and is equal to $1/M_p$, where M_p represents the number of villages in the pth PSU ($p = 1, \ldots, 40$ in the sample). For most exposed LC3s in Uganda, there is only one exposed LC1 per PSU. Where there is more than one exposed village per PSU, the SSU was drawn with a random number from the list of those villages with more than 100 households according to the 1991 census. Whether a community selected in the sample had been properly classified as exposed or nonexposed was then verified at the site. Two communities were replaced because they were fishing villages with little banana production.

Selection of Households within a Community

A constant sample size was maintained for two reasons. First, as mentioned above, there was no information on variation in underlying population parameters to guide optimal sample allocation. The second reason was to maintain comparable workloads among enumerators, who resided in the communities for 1 year to complete the survey. The sample villages were visited, and a current list of households in each village was requested from the chairman of the LC1. The total number of households selected per village was 20. The probability of selection (or a sampling fraction) of a household varies by village and is equal to $20/H_s$, where H_s is the number of households in village s ($s = 1, \ldots, 40$ in the sample). If there was an order in the list of households, random numbers were used for selection. Otherwise, a random start with systematic random sampling from the list was employed.

Unit of Observation

The basic unit of observation for the sample survey is the farm household. A farm household is defined according to the culture of which the household is a part, and includes female-headed and child-headed (orphaned) households, as well as male-headed households with more than one wife. The total sample size surveyed comprises 800 households (540 in Uganda and 260 in Tanzania). In some analyses conducted in this report, the plot or the banana cultivar is the unit of observation.

Weights

The overall probability of selecting a household in the subsample (denoted as PSH) is a unique number, defined as the product of the sampling fractions at each level: PSH =

$[(n_i/N_i) \times (1/M_p) \times (20/H_s)]$. For descriptive analysis, survey weights (w) were calculated for each household as the inverse PSH.

Survey Instruments

A set of 10 structured, pretested questionnaires (schedules) were used as instruments for data collection, each designed to address a different aspect of the study. Six of the instruments were single-visit: household, banana plot, banana cultivar, labor, expenditure, and income. Data for the general plot schedule (the seventh schedule) were collected three times, to capture production seasonality. Expenditure, labor, and production and income data (the remaining three schedules) were collected monthly.

The format and structure of the banana plot and banana cultivar instruments, as well as the banana management and social capital instruments, depart in some ways from the more typical household and plot surveys often conducted in studies of technology use by smallholder farmers. These are described briefly below.

Banana Plot and Cultivar Schedules

The banana plot schedule records

1. the primary decisionmaker for banana production, consumption and sales, disaggregated by gender and category of responsibility, with frequency of exposure to extension and radio information;
2. an inventory of all banana varieties on the plot, with mat counts and average number of plants per mat, by genomic group and clone set (recorded later); and
3. perceptions of incidence and severity of disease (black Sigatoka and *Fusarium* wilt) and pest (weevils) pressure.

Nematodes were omitted because they were not observable by farmers. Farmers were asked the frequency of occurrence of the disease or pest in all years growing bananas on the plot, along with the age of the plot. Triangular yield distributions in the presence and absence of the disease or pest were elicited by cultivar (Hardaker, Huirne, and Anderson 1997).

This cultivar schedule records selected phenotypic or morphological characteristics[1] (as identified by farmers) for each banana cultivar recorded in the banana plot schedule. Production and consumption attributes identified as important to farmers in previous field research in Uganda were listed. Banana production and consumption decisionmakers were then asked to rate the importance of each attribute and the extent to which each cultivar provides the attribute. Color photographs and drawings were employed to assist respondents. The same information was elicited for banana varieties that farmers had previously grown or that were grown in their village but not by them. This schedule also records information about farmer-to-farmer transfers of banana planting material and acquisition, by cultivar, with questions on willingness to pay or willingness to accept compensation for planting material. Characteristics of market sales of banana bunches (amount, farm-gate and market prices, and distance) were entered.

Banana Management Schedules

The banana management schedules elicited information on the farmer's management of the natural resources in their banana plantations, as well as sanitation practices (mat management) for pest and disease control, including use and awareness of recommended practices and sources of information about management practice. The extent

[1] Morphological characteristics are the observable physical features that help farmers distinguish one banana cultivar from another, such as bunch size and position, finger size and position, maturation period, and color of the stem of the banana plant.

of use for the organic fertilizers for a one-production cycle (January to December 2003) was measured by counting the number of mats under each type of organic fertilizer. To minimize the measurement error, the interview was conducted in the plot, and the farmer showed the enumerator parts of the plantation that were treated with the organic fertilizers as the enumerator counted mats in the area. For mat management practices, color photographs were used to enhance the farmer's recognition of the practice being surveyed.

Social Capital Schedules

Social capital data were collected through discussions with key informants and the sampled households. Key informants (the local leaders and village elders) were interviewed about village social homogeneity (in terms of ethnicity and religious affiliation) and formal and informal organizations.

The schedule on associations recorded information on household membership; the level of household participation; major activities of associations; benefits to the members; and composition, function, and leadership quality of associations. These were measured following the work of Narayan (1997). Social networks, trust, and solidarity, as well as historical rules and regulations, were measured in other instruments as part of the thesis by Katungi (2006) but are not presented here.

References

Chandler, S. 1995. The nutritional value of bananas. In *Bananas and plantains,* ed. S. R. Gowen. London: Chapman & Hall.

De Groote, H. 1996. Optimal survey design in rural data collection in developing countries. *Quarterly Journal of International Agriculture* 35 (2): 163–175.

Hansen, M. H., W. N. Hurwitz, and W. G. Madow. 1953. *Sample survey methods and theory.* New York: John Wiley and Sons.

Hardaker, J., R. Huirne, and J. Anderson. 1997. *Coping with risk in agriculture.* Wallingford, U.K.: CAB International.

Karamura, D. A., and E. Karamura. 1994. *A provisional checklist of banana cultivars in Uganda.* Kampala, Uganda: Kawanda Agricultural Research Institute, Uganda National Banana Research Programme, National Agricultural Research Organization.

Katungi, E. 2006. Social capital and technology adoption on small farms: A case of banana production technology in Uganda. Ph.D. thesis, University of Pretoria, Pretoria, Republic of South Africa.

Narayan, D. 1997. *Voices of the poor: Poverty and social capital in Tanzania.* Environmentally and Sustainable Development Studies and Monographs Series 20. Washington, D.C.: World Bank.

Speijer, P. R., C. S. Gold, E. B. Karamura, and J. N. Kashaija. 1994. Banana weevil and nematode distribution patterns in highland banana systems in Uganda: Preliminary results from a diagnostic survey. In *African Crop Science Conference proceedings,* ed. E. Adipala, M. A. Bekunda, J. S. Tenywa, M. W. Ogenga-Latigo, and J. O. Mugah. June 14–18, 1993, Kampala, Uganda. Kampala: African Crop Science Society.

Tushemereirwe, W. K., D. A. Karamura, H. Sali, D. Bwamiki, I. Kashaija, C. Nankinga, F. Bagamba, A. Kangire, and R. Sebuliba. 2001. Bananas (*Musa* spp.). In *Agriculture in Uganda.* Vol. 11, *Crops,* ed. J. Mukiibi. Kampala, Uganda, and Wageningen, The Netherlands: Fountain Publishers, Technical Centre for Agricultural and Rural Cooperation, and National Agricultural Research Organization.

APPENDIX E

Village Social Structure in Uganda

Enid Katungi

The overwhelming significance of farmer-to-farmer networks to the exchange of planting material and information about recommended management practices underscores the role of village social structures in determining the use of banana technologies. This appendix briefly discusses the current state of local social structure in the selected banana producing areas of Uganda, focusing specifically on village homogeneity (in terms of ethnicity and religion) and the participation in associations by agricultural households. There is evidence that information sharing among agricultural households depends on the characteristics of the social structures in which they live, and that some village norms are more conducive to sharing information and technology transfer or adoption. More extensive information about social capital and household interconnectedness is found in Katungi (2006).

A high degree of social homogeneity, expressed in terms of domination by a single ethnic group,[1] is apparent in Uganda (Table E.1). Highland villages were found to be more homogenous (more than 90 percent in one ethnic group) than lowland villages. Only 4 out of 20 villages surveyed had less than 50 percent of households from a single ethnic group, and all were found in the lowlands. According to key informants, the ethnic homogeneity in these villages was affected by the importation of labor from other regions to work in commercial crops (coffee and cotton), sugar factories, or railway construction during the colonial period.

Villages are far less homogeneous in terms of religious groups. Like ethnic group homogeneity, religious homogeneity is greater in the highlands (67 percent in a single religion) than in the lowlands (58 percent). In Uganda, religious wars were fought even before the advent of multiparty politics, although multiparty politics reinforced these divisions, which are still evident 18 years after political change. About 30 religious-based community organizations distributed in 13 villages were found. How these structures influence technology diffusion de-

[1] The concept of an ethnic group cuts across various forms of social organization, including tribe and kingdom, but conveys more of a shared sense of territory and a link to an ecology or food culture than does the concept of language group. Some ethnic groups have lost the use of their local languages and yet still remain distinct ethnic groups with a unique ecology and food culture. The term "ethnic group" has no colonial connotations and represents a level of institutions and social organization within and at times across the nation-state (Pablo Eyzaguirre, IPGRI, personal communication, June 10, 2005). Here, the concept of ethnicity is used here to refer to a social group of people with shared tribal affiliation based on patrimonial lineage. Information about ethnicity and religion was obtained through key informant interviews, and households were classified according to the group membership of the household head. Note that it is the ethnic and/or religious group of the household head that is used to classify households.

Table E.1 Selected indicators of social homogeneity in rural villages, Uganda

Indicator	Elevation		All
	Low	High	
Percentage of households with same ethnicity	68.5***	95.1***	76.09
Percentage of households with same religion	58.2***	67.00***	59.50
Number of ethnic groups	5.4***	2.00***	4.57
Number of religious groups	5.0***	3.8***	4.59

Note: *** indicates statistical significance at the 1 percent level for differences.

pends on which of these attributes influences the nature and content of social interactions among households.

Most of the villages in the banana producing areas foster active social organization. More than 250 associations were identified in 20 villages, categorized in Table E.2 as (1) burial societies, (2) associations with economic orientation (formal and informal credit, farm production, and trade-based associations), (3) faith-based associations, and (4) culturally oriented associations.[2]

There are significantly more economically oriented associations per village in high-altitude areas compared to low altitude (Table E.2). The number of informal credit associations per village was higher in high-altitude areas, whereas the number of formal credit associations was higher in low-altitude areas. Households at high elevations are relatively more constrained in land, which implies that they are more likely to lack collateral for use in formal institutions and hence derive more benefits from informal credit associations. Statistical differences in the number of informal credit associations were also found according to distance from paved roads, but it is interesting that similar differences were not apparent in the number of formal credit associations. Associations dealing in agriculture are more prevalent in the lowlands because of the relatively high level of recent intervention in agriculture by NGOs and NARO, especially in the Central Region. Most of the agricultural associations were initiated by an external influence.

By far the greatest number of respondents reported that decisionmaking processes in associations are participatory rather than consultative or dictatorial. Most associations drew membership from within the village geographical boundaries. Credit associations are more likely to cut across ethnic groups. Associations are not homogeneous with respect to gender.

Among all households surveyed, 76 percent belonged to at least one association (Table E.3). Social organizations were more popular than economic associations. Most households belong to these for emotional and material support or to conform to the community expectations. Burial societies, the most common social association, are essentially a means of pooling resources to organize and pay for unexpected expenses, such as funerals. There is in general only one burial society in a village, which has both rich and poor members. Although there are no entry fees, every member is expected to contribute money and other resources toward funerals.

Despite the potential benefits of economic associations and their importance in

[2] In rural areas, functions (social or economic) of an association may overlap. Here, categorization of associations is based on the dominant activity of the association, and hence the groupings are not mutually exclusive.

Table E.2 Average number of associations per village, by type of association, Uganda

Association type	Elevation Low	High	All
Total per village	11.69***	15.65***	13.06
Burial societies	2.27***	1.21***	1.88
Economically oriented associations	8.16***	9.72***	8.70
Informal credit	2.04***	6.23***	3.56
Formal credit	1.87***	0.43***	1.44
Agricultural	4.00***	2.23***	3.46
Trade-based	0.38***	0.83***	0.52
Culture-based	0.34***	0.60***	0.45
Faith-based	2.5**	3.59**	2.69

Note: *** and ** indicate statistical significance at the 1 and 5 percent levels, respectively.

Table E.3. Percent of households belonging to associations in Uganda

Association type	Elevation Low	High	All
At least one association	70.63***	93.68***	76.38
Burial societies	47.55***	92.63***	58.79
Economically oriented associations	41.26***	60.00***	45.9
Informal credit	12.94***	48.42***	21.78
Formal credit	16.44***	4.21***	13.36
Agricultural	3.15	4.21	3.41
Trade-based	20.63	13.68	18.90
Culture-based	2.80	4.21	3.15
Faith-based	6.64***	23.16***	10.76

Note: *** indicates statistical significance at the 1 percent level.

overcoming market imperfections, only 46 percent of households surveyed were members of economic associations, raising important questions about the determinants of participation. Participation of individuals in most types of associations, except in formal credit associations, is significantly greater in the highlands than the lowlands.

Reference

Katungi, E. 2006. Social capital and technology adoption on small farms: A case of banana production technology in Uganda. Ph.D. thesis, University of Pretoria, Pretoria, Republic of South Africa.

List of Principal Authors and Contributors

Pamela Nahamya Abodi is a senior research assistant at the IFPRI Kampala, Uganda, office. She has worked with IFPRI since November 2000 on a number of projects that focused on policies for improved land management in Uganda and banana-specific studies in the field of biotechnology economic impact assessment in East Africa. She has been working with research teams involved in survey design, data collection and processing for analysis, and report writing. She holds an M.S. in crop biotechnology (2001) and a B.S. in agriculture (1998) from Makerere University, Kampala.

Fredrick Bagamba has worked as a research scientist with NARO for the past ten years. He is currently a Ph.D. student with the Development Economics Group, Wageningen University, Wageningen, The Netherlands. His current research focuses on biophysical and economic aspects influencing productivity of banana production systems in Uganda, with a particular focus on natural resource endowments and market access in banana production. Fredrick obtained an M.S. in agriculture from Makerere University, Kampala, Uganda (1995). The subject of his M.S. thesis was resource allocation efficiency in a banana-based cropping system in Uganda.

Mgenzi Said Ramadhan Byabachwezi is a research officer at Maruku Research and Development Institute, Bukoba, Tanzania. He earned his M.S. in ecological agriculture at Wageningen University, Wageningen, The Netherlands (2000). He also has a B.S. in agriculture, obtained at Sokoine University of Agriculture, Morogoro, Tanzania (1991). He has worked on collaborative research activities with INIBAP, the Association for Strengthening Agricultural Research in Eastern and Central Africa's FOODNET program, and IFPRI. His specialization is plant protection, mainly in bananas. He has also been working on banana taxonomic activities. He was an analyst for the research summarized in this report.

Svetlana Edmeades has been an economist at the Agriculture and Rural Development Department at the World Bank, Washington, D.C., since September 2006. Before joining the World Bank she was a postdoctoral fellow at IFPRI. She earned her Ph.D. in economics from North Carolina State University (2003), specializing in the fields of development and natural resources/environmental economics. Her dissertation focused on the development of a conceptual framework for the analysis of variety demand in semisubsistence agriculture, where markets are imperfect and variety characteristics are important factors in farmers' decisions about growing staple crops. Her dissertation work was part of the research summarized in this report. Svetlana also holds an M.S. in economics from the University of Waikato, New Zealand (1998) and a B.S. in international relations from the University of the Americas, Mexico City (1995).

Charles A. Eledu joined IPGRI-INIBAP in June 2000 as a geographic information systems specialist to work for INIBAP's banana baseline information project in East and southern Africa, based in Kampala, Uganda. Charles has over 10 years of R&D experience, having previously worked with the Uganda Agriculture Ministry, Makerere University Institute for Social Research, and the International Centre for Tropical Agriculture as agricultural officer, research associate, and geographic information systems research associate, respectively. He holds an M.S. in soil survey (application of soil survey information for natural resources management) from the International Institute for Geo-Information Science and Earth Observation, Enschede, The Netherlands, a B.S. in agriculture (management and extension) from Makerere University, Kampala, Uganda, and several other professional and technical postgraduate qualifications.

Robert Kalyebara is a senior economist with NARO. He received a Ph.D. in agricultural economics from Ohio State University in 2001 and has worked in research for 19 years, mainly on crops. His main areas of specialization are impact assessment/research evaluation and monitoring and evaluation.

Deborah Karamura joined IPGRI-INIBAP in January 2000 as a *Musa* genetic resource specialist, responsible for INIBAP's in situ banana conservation project in East and southern Africa, based in Kampala, Uganda. Applying her wide experience in banana taxonomy and nomenclature, particularly of highland bananas, Deborah has developed and tested methodologies for the characterization and conservation of *Musa* diversity and created innovative approaches for mobilizing and supporting farming communities in East Africa in the quest to conserve banana diversity on farm. Her research focuses on the determination and documentation of cultivar diversity in traditional banana-based systems; the identification of genetic erosion factors; the documentation of traditional management approaches, with the aim of integrating them with research-derived technologies through scientific methodologies; and the use of banana diversity to improve rural livelihoods as an on-farm conservation strategy.

Enid Katungi is an economist with NARO and a Ph.D. student of agricultural economics at the University of Pretoria, South Africa. Her dissertation focuses on social capital and technology adoption on small farms, exploring the impact of community associations, networks, and other social characteristics on information diffusion and the use of agricultural technologies. The dissertation also employs an actor-centered approach and analyzes the levels and determinants of social capital among the agricultural households in the banana growing regions of Uganda. Currently she is also working on the role of participatory technology dissemination methods in technology adoption, focusing on banana technologies in Uganda. Enid obtained her M.S. in agricultural economics (1998) and a B.S. in agriculture science (1992) both from Makerere University, Kampala, Uganda.

Enoch Mutebi Kikulwe is an economist with the National Banana Research Programme of NARO. He obtained an M.S. in agricultural economics from Makerere University, Kampala, Uganda. He has worked on several research projects in Uganda, which involved survey design, data collection, and data analysis. Enoch has worked as a data manager for this project and is a doctoral student at the University of Wageningen, Wageningen, The Netherlands.

Charles Machethe is professor in the Department of Agricultural Economics, Extension and Rural Development, the University of Pretoria, Pretoria, South Africa. His major areas of research are in agricultural and rural development, rural finance, economics of smallholder irrigation, and farm-nonfarm linkages. He has a Ph.D. and M.S. in agricultural economics from Michigan State University, East Lansing, and an M.S. in agriculture from the University of the North, Limpopo, South Africa.

Josephine Mary Namaganda is a senior research officer/plant nematologist with NARO. She earned her Ph.D. in plant nematology from the University of Reading, U.K., in 1996. Her research interests are banana nematology and tissue culture biosafety, including environmental risk assessment.

Jackson M. Nkuba is an agricultural economist and a researcher at the Maruku Agricultural Research and Development Institute, Bukoba, Tanzania. Currently he is a Ph.D. student at the Sokoine University of Agriculture, Morogoro, Tanzania. He obtained his M.S. in agricultural economics from the University of Queensland, Brisbane, Australia (1996), and his B.S. in agricultural economics from Sokoine (1990). He has experience in research management, participatory approaches, and monitoring and evaluation of projects, and is interested in environmental economics. His doctoral research is about the adoption and economic impacts of new banana cultivars on the livelihoods of people in Kagera Region, Tanzania.

Kephas Nowakunda works for NBRP. He is currently pursuing a Ph.D. in food processing at the National University of Ireland, Galway (2003–07). Kephas holds an M.Sc. in agriculture (postharvest technology, 2001) and a B.S. in food science and technology (1994), both awarded by Makerere University, Kampala. He joined NBRP as a postharvest scientist in 1995, responsible for implementing multilocation evaluations of new banana germplasm. Kephas has also been involved in international and national projects dealing with postharvest handling systems, value addition, quality control, and management of quality control systems and marketing. He was awarded a European Union Scholarship to train in novel food processing and quality control systems and management in Valencia, Spain (2003). He has also undertaken study visits in several countries, including Peru and Bolivia, to study the marketing and quality control systems developed under projects funded by USAID/INIBAP. Following completion of his doctorate, he will be responsible for postharvest, quality, and market-testing studies. Before joining NBRP, Kephas worked as a quality controller for the Coffee Marketing Board, and later as a quality control systems coordinator at the Coca-Cola Century Bottling Company in Kampala, Uganda.

Ingrid Rhinehart is an economist who works with IFPRI as a private consultant. Her research interests include spatial analysis of agricultural and poverty issues. She specializes in spatial econometrics and geographic information systems. Ingrid holds a B.A. in economics from Oberlin College, Oberlin, Ohio, and an M.S. in agricultural economics from Cornell University, Ithaca, New York, where she concentrated on human-wildlife conflict issues in East Africa.

Ruerd Ruben is associate professor in the development economics group at Wageningen University, Wageningen, The Netherlands. He holds a Ph.D. in development economics from Free University, Amsterdam, and is chair of the Development Studies group at Radboud University, Nijmegen, The Netherlands. His research activities are centered on economic policies for sustainable agriculture and incentives for rural development in less-favored areas. He conducted extensive field research in Central America, East and West Africa, and southern China, focusing on the dynamics of land, labor, markets, and supply-chain linkages between smallholders and retail outlets.

Mariana Rufino is currently a Ph.D. student with Plant Research International, Wageningen University, Wageningnen, The Netherlands. She did her M.S. with the Plant Sciences Department, Production Systems Group, Wageningen University. The subject of her thesis is on-farm analysis of nematode infestation and soil fertility as constraints to the productivity of banana-based production systems in Uganda.

Melinda Smale leads a research program on economics and genetic resources at IFPRI and IPGRI. Her research emphasizes the development of methods to assess the value of crop biodiversity and the identification of policies to enhance the utilization and management of crop genetic resources, particularly in developing economies. From 1994 to 2000, in Mexico, Melinda worked on crop genetic resources and technology adoption issues with CIMMYT. She conducted research in Malawi from 1989 to 1993 on hybrid maize adoption by smallholder farmers and maize research impacts, also with CIMMYT. During the 1980s, she worked in Pakistan, Somalia, Mauritania, and Niger on short-term assignments for CIMMYT, Chemonics International, Volunteers in Technical Assistance, and USAID. She obtained her Ph.D. in agricultural economics from the University of Maryland, College Park, her M.S. in agricultural economics from the University of Wisconsin, Madison, and her M.A. specializing in Africa area studies from The Johns Hopkins School of Advanced International Studies, Washington, D.C.

David Talengera is a researcher with the biotechnology unit of the National Banana Research Programme and a Ph.D. student at the University of Pretoria, South Africa. He studied cell and plant tissue culture under NARO and IITA's joint banana breeding project between 1995 and 2005. He obtained his M.Sc. and a B.Sc. in agriculture from Makerere University, Kampala, in 1994 and 1990, respectively.

Wilberforce Kateera Tushemereirwe is program leader, NBRP, NARO, based at KARI. Since 1997, he has led a national team in R&D of the banana crop. For the past 4 years he has developed and implemented successful projects with the goal of building crop biotechnology capacity, with a team of young scientists eager to learn and apply biotechnology. Dr. Tushemereirwe also has extensive experience and expertise in formulating and managing research projects, using participatory approaches, communicating results to stakeholders, and transferring technologies to the private and commercial sectors. He holds a Ph.D. in plant pathology and an M.S. in botany from the University of Reading, England.

Stanley Wood, senior scientist/program head, joined IFPRI in 1995 and until 1997 was outposted to the Centro Internacional de Agricultura Tropical in Cali, Colombia, working on impact assessment studies for regional agricultural research. Since then he has been based at IFPRI headquarters, where he initially lead IFPRI's research on spatial analysis in a policy context and now co-leads IFPRI's new Global Change program. Before joining IFPRI, he served as an independent consultant to multilateral and bilateral aid organizations on natural resource and agricultural systems modeling. He earned a B.E. in civil engineering from the University of Sheffield, Sheffield, England, an M.S. in water resources development from the University of Birmingham, Birmingham, England, and an M.S. in agricultural development and a Ph.D. in economics from the University of London.

Made in United States
Troutdale, OR
02/28/2026

47317304R00117